Marie Curie

Significant Figures in World History

Charles Darwin: A Reference Guide to His Life and Works,
by J. David Archibald, 2019.

Leonardo da Vinci: A Reference Guide to His Life and Works,
by Allison Lee Palmer, 2019.

Michelangelo: A Reference Guide to His Life and Works,
by Lilian H. Zirpolo, 2020.

Robert E. Lee: A Reference Guide to His Life and Works,
by James I. Robertson Jr., 2019.

John F. Kennedy: A Reference Guide to His Life and Works,
by Ian James Bickerton, 2019.

Florence Nightingale: A Reference Guide to Her Life and Works,
by Lynn McDonald, 2020.

Napoléon Bonaparte: A Reference Guide to His Life and Works,
by Joshua Meeks, 2020.

Nelson Mandela: A Reference Guide to His Life and Works,
by Aran S. MacKinnon, 2020.

Winston Churchill: A Reference Guide to His Life and Works,
by Christopher Catherwood, 2020.

Catherine the Great: A Reference Guide to Her Life and Works,
by Alexander Kamenskii, 2020.

Marie Curie: A Reference Guide to Her Life and Works,
by Marilyn Ogilvie, 2021.

Marie Curie

A Reference Guide to Her Life and Works

Marilyn Ogilvie

ROWMAN & LITTLEFIELD
Lanham • Boulder • New York • London

Published by Rowman & Littlefield
An imprint of The Rowman & Littlefield Publishing Group, Inc.
4501 Forbes Boulevard, Suite 200, Lanham, Maryland 20706
www.rowman.com

6 Tinworth Street, London, SE11 5AL, United Kingdom

Copyright © 2021 by The Rowman & Littlefield Publishing Group, Inc.

All rights reserved. No part of this book may be reproduced in any form or by any electronic or mechanical means, including information storage and retrieval systems, without written permission from the publisher, except by a reviewer who may quote passages in a review.

British Library Cataloguing in Publication Information Available

Library of Congress Cataloging-in-Publication Data

Names: Ogilvie, Marilyn Bailey, author.
Title: Marie Curie : a reference guide to her life and works / Marilyn Ogilvie.
Description: Lanham : Rowman & Littlefield, [2020] | Series: Significant figures in world history | Includes bibliographical references and index. | Summary: "This encyclopedia examines Marie Curie's life and contributions"—Provided by publisher.
Identifiers: LCCN 2020035844 (print) | LCCN 2020035845 (ebook) | ISBN 9781538130018 (hardcover) | ISBN 9781538130025 (ebook)
Subjects: LCSH: Curie, Marie, 1867–1934. | Chemists—France—Biography. | Women chemists—France—Biography. | Physicists—France—Biography. | Women physicists—France—Biography.
Classification: LCC QD22.C8 O355 2020 (print) | LCC QD22.C8 (ebook) | DDC 540.92 [B]—dc23
LC record available at https://lccn.loc.gov/2020035844
LC ebook record available at https://lccn.loc.gov/2020035845

∞™ The paper used in this publication meets the minimum requirements of American National Standard for Information Sciences—Permanence of Paper for Printed Library Materials, ANSI/NISO Z39.48-1992.

Contents

Preface vii

Chronology ix

Introduction 1

ENTRIES A–Z 11

Bibliography 95

Index 109

About the Author 115

Preface

Throughout history, most of the people involved in the scientific enterprise were men. This observation has typically been explained in one of two ways: either women were unsuited by their nature to be scientists, or their training or education made it difficult for them to succeed in scientific fields. Aristotle (384–322 BCE), whose influence on future science cannot be overstated, blamed women's "intellectual disability" on the fact that "the female, is female on account of an inability of a sort *viz*, it lacks the power to concoct semen. . . . Femaleness should be considered a 'deformity,' though one which occurs in the ordinary course of nature" (*De generatione animalium* I.20, 4.60). Although nurture advocates also were found throughout history, they were in the minority. For instance, Aristotle's teacher, Plato (424/423 BCE–348/347 BCE), was more sympathetic to educating women than his student. He noted that it would be a waste of talent to exclude 50 percent of the population in intellectual pursuits. If women received the same kind of education as men, they would perform just as well.

In spite of gender barriers, women interested in scientific questions throughout history managed to make contributions to science, although often their impact was more in the factual realm than the theoretical. In order to compete in the male domain of science, women developed various strategies, including impersonating men; choosing fields that could be considered "women's work" such as home economics, science popularization, and scientific illustration; and collaborating with a colleague or male relative. This latter strategy, collaborating with a successful male relative, Pierre Curie, helped Marie Curie succeed. Still, those who preferred to think that women could not think creatively argued that it was Pierre who was responsible for the major ideas and that Marie was merely an efficient helper and implementer. Although it is correct that Marie's most scientifically creative years were those in which she and Pierre shared ideas, the basic hypotheses that guided their future course of investigation into the nature of radioactivity were hers.

This encyclopedia examines Marie Curie's life and contributions in several different ways. The chronology provides a thumbnail sketch of events in Curie's life, including her personal experiences, education, and publications. The introduction provides a brief look at her life. The body of this work consists of alphabetical entries on people, ideas, institutions, places, and publications important in making Curie an important scientist. The final section of the book is a bibliography of both primary and selected secondary sources.

My career as a historian of science began with a dissertation on the pre-Darwinian evolutionist Robert Chambers and his work *Vestiges of the Natural History of Creation*. I became interested in women in science while teaching a survey course in the history of science. I assigned a paper to my students, explaining that they could choose their topic as long as it was in the proper chronological period, but they must clear their topic with me. After class, two young women appeared and

PREFACE

announced that they planned to write on a woman in science. The next class period they approached me again, this time with the news that they could only find one woman in science, and that woman was, of course, Marie Curie. I was of little help! Their interest and my ignorance set me on a search for other women in science. Others in the history of science community also were becoming interested in the subject. We found that actually women had been engaged in the scientific enterprise from antiquity to the present. My search led to two biographical dictionaries of women in science (one with Joy Harvey), numerous articles, and several full-length biographies. One of these biographies was on Marie Curie, who had started my quest. When Rowman & Littlefield asked me to work on this encyclopedia, I felt that I had come full circle.

In order to make this book a useful reference tool, I have cross-referenced entries in the encyclopedia section. Within individual entries, terms that have their own entries are in boldface type the first time they appear. Related terms that do not appear in the entry are indicated by "see also" references.

Chronology

1860 Bronislawa Boguska (Marie Curie's [née Maria Sklodowska] mother) marries Maria's father, Wladyslaw Sklodowski, and she becomes headmistress of a private school for girls in Warsaw, the Freta Street school. The couple moves to an apartment furnished by the school and adjacent to it. Their five children are born in this house.

1867 7 November: Maria Salomea Sklodowska is born the fifth of five children of Bronislawa and Wladyslaw Sklodowski in Warsaw, Poland.

1868 The Sklodowsi family moves to the Nowolipki apartments provided by Wladyslaw's school.

1871–1873 When Maria is four, her mother's tuberculosis worsens and she leaves home for a spa for two years, leaving the children with their father.

1873 Wladyslaw Sklodowski is dismissed from his position as assistant director of the Nowolipki gymnasium, the family loses its living quarters, and they turn their new home into a boarding school.

1874 Maria's oldest sibling, Zofia (Zosia), dies of typhus.

1876 Maria attends the Sikorska School.

1878 9 May: Maria's mother, Bronislawa, dies of tuberculosis. Maria enrolls in the Gymnasium Number Three.

1883 Maria graduates from secondary school (Gymnasium Number Three) with a coveted gold medal.

1886 Maria begins a job as governess with the Zorawskis.

1889 Maria leaves her job with the Zorawskis.

1891 5 November: Maria registers as Marie (the French form of Maria) Sklodowska as a student at the Sorbonne.

1893 June: Marie earns a first-place degree in physics (*licenciée ès sciences physiques*).

1894 July: Marie earns a second-place degree in mathematics (*licenciée ès sciences mathématiques*). While she is finishing this mathematics degree, she is hired by the Society for the Encouragement of National Industry, an organization formed to promote French science. She is to determine the magnetic properties of different types of steel. As she is hunting for laboratory space for this project, she meets Pierre Curie.

CHRONOLOGY

1895 26 July: Marie marries Pierre Curie. She assists Pierre in preparing his teaching courses at the School of Industrial Physics and Chemistry.

1897 Fall: Marie completes her first paper on the magnetic properties of various tempered steels. Pierre and the pregnant Marie take a cycling trip to Brest. They cut the trip short, and Marie gives birth to Irène on 12 September 1897. Pierre's mother dies of cancer two weeks after Irène is born.

1898 18 July: Marie and Pierre announce the discovery of polonium. **26 July:** Marie and Pierre, with Gustave Bémont, announce the discovery of radium. **12 September:** Marie introduces the term *radioactivity* in a published article. **12 December:** Marie is awarded the 3,800-franc Prix Gegner for her work on the magnetic properties of steel as well as her work on radioactivity.

1900–1906 Marie teaches at the École Normale Supérieure de Jeunes Filles in Sèvres.

1903 11 May: Marie's doctoral thesis is approved by the Sorbonne's dean of the faculty. **June:** Marie defends her dissertation. **Summer:** The thesis is published in England in Crooke's *Chemical News* and in September in France in the *Annales de Physique et de Chimie*. **August:** Marie loses a baby born prematurely. **December:** Nobel Prize in Physics is shared between Marie and Pierre Curie and Henri Becquerel. The Curies' prize is for "their joint researches on the radiation phenomenon discovered by Professor Henri Becquerel."

1904 6 December: Ève Curie is born.

1906 19 April: Pierre dies in an accident. **5 November:** Marie becomes the first woman professor at the Sorbonne.

1911 23 January: Marie fails to be elected to the French Academy of Science and loses to Edouard Branley. **December:** Marie is awarded the Nobel Prize in Chemistry.

1914 The Radium Institute in Paris is completed.

1914–1919 Marie operates a mobile X-ray unit during World War I.

1921 May–June: Marie with her two daughters visits the United States to receive a gram of radium collected by Marie Meloney from wealthy women in the United States.

1929 16 October–8 November: Marie again visits the United States, this time to receive a gram of radium for her new institute in Poland.

1934 4 July: Marie Curie dies of aplastic anemia. The French government abandons its plans for a national funeral in accordance with her wishes and that of her family. **5 July:** Her body is brought to Paris. **6 July:** She is buried in a simple ceremony without a civil or religious service in a plain oak coffin in Sceaux in the same grave where Pierre had been placed.

1995 20 April: The remains of both Pierre and Marie are transferred to the Pantheon in Paris, where they were reburied. Marie Curie is the first woman so honored for her own accomplishments.

Introduction

The name of Maria (Marie) Sklodowska Curie is well known to both scientists and laypersons and invariably surfaces when people are asked to name an important woman scientist. However, it is not only her unique status as a female scientist that sets her apart, although certainly this is important, but as one in a Pantheon of late 19th- and early 20th-century scientists who have overturned traditional ideas about the nature of the universe. Curie's part in this new scientific revolution was pivotal. Her interaction with the ideas of contemporary scientists and theirs with hers was vital in the construction of this new scientific universe. She made one of the most important theoretical breakthroughs of the 20th century when she postulated that radiation was an atomic rather than a chemical property. Rather than being an ordinary chemical reaction where light or heat is produced, she demonstrated that radiation issued from the atom itself and was proportional to the amount of the radioactive substance being measured. Her more prosaic but better-known contributions included being the first to use the term *radioactivity* and her long search that culminated in the isolation of two new elements, polonium and radium.

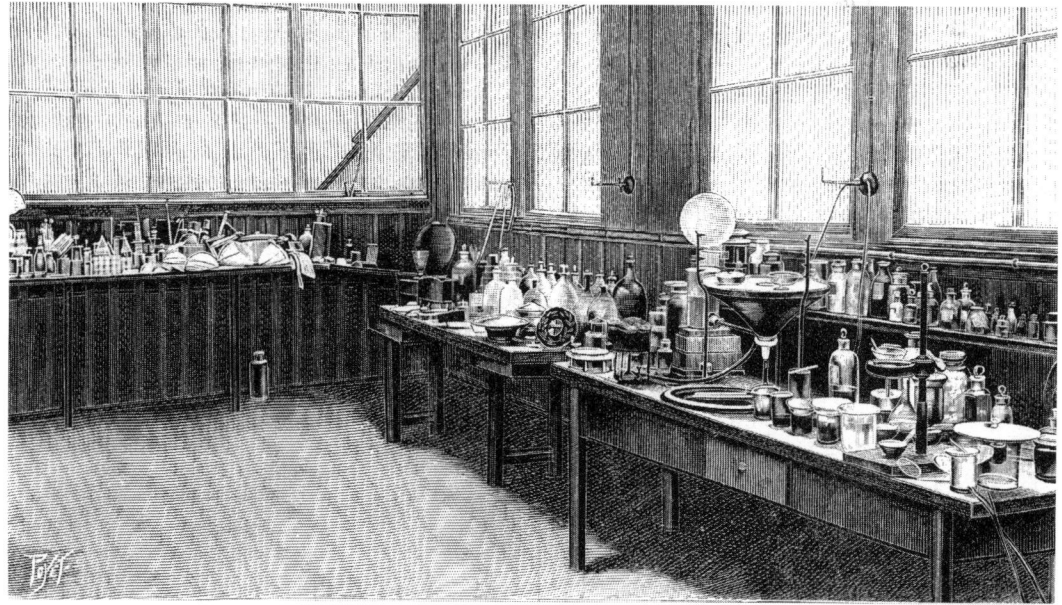

Marie and Pierre Curie's laboratory

INTRODUCTION

To succeed in science in the late 19th century, Curie had to create a place for herself in an almost exclusively masculine world of laboratory science. Her success as a woman scientist was the product not only of her creativity and perseverance but also of the relationships that she developed with her male colleagues. Some of these relationships were positive, especially her partnership with her husband, Pierre; others, including an affair with the physicist Paul Langevin, threatened to disrupt her career. Her success as a woman in science paved the way for subsequent women scientists. As the first woman to receive the Nobel Prize (1903, physics, shared with Pierre and Henri Becquerel) and the first person to receive two Nobel Prizes (second Nobel, chemistry, 1911), she was the recipient of the respect not only of the scientific community but of the population at large. In spite of her achievements, Curie had to fight the assumption that she had been merely Pierre's assistant, since women were not expected to make theoretical contributions to science.

Maria Sklodowska's childhood in a Poland dominated by Russia was influenced by the political climate. Poland's government had long been inefficient, chaotic, and corrupt, making it ripe to be overrun by efficient, stronger neighbors—Austria, Russia, and Prussia. In their drive to increase their own territories, these three countries partitioned Poland among themselves at different times. Poland hoped to find a benefactor in Napoleon, who after he had defeated Austria and Prussia helped the country raise its own army. He created the Duchy of Warsaw and raised the spirits of the Poles, who anticipated regaining control of their country. However, after Napoleon's disastrous defeat by Russia (1811–1812), that country stepped in and replaced Napoleon's Duchy of Warsaw with the Kingdom of Poland, connected to Russia by a union with Russia's tsar. The tsar became the king of Poland, which had its own constitution, parliament, army, and treasury. The remaining territories were united under Prussian rule. After several unsuccessful uprisings, the Polish resistance failed, and many concessions the Poles had extracted from the Russians were taken away. The hope of defeating Russia by military action foiled, civil disobedience seemed to be the most effective means of protest. As the peasants in both Russia and Poland demanded rights, the Russian tsar was forced to court them in order to ensure both countries ran smoothly. The Russian serfs were enfranchised in 1861, and the Poles demanded the same freedoms in 1864. However, the enfranchisement did not have the desired effect in Poland. The peasants became active members of the National Polish Community, which was a hotbed of rebellious Poles. Polish attempts to overthrown Russian rule nevertheless failed, and the subsequent Russian oppression was brutal.

Maria Sklodowska Curie was born into this political atmosphere. Her father, Wladyslaw Sklodowski, had obtained a scientific education in Russia. When he returned to Warsaw to teach physics, he met Bronislawa Boguska, a principal of a girls' boarding school. Both Wladyslaw and Bronislawa were members of the minor nobility, but because of the political situation neither had sufficient money to live comfortably, causing them to economize drastically. The couple had five children: Zofia, known as Zosia (b. 1862); Jozef (b. 1863); Bronislawa, named for her mother and known as Bronia (b. 1865); Helena, known as Hela (b. 1866); and Maria, known as Manya (1867). During the first eight years of their marriage, when their family was growing, the Sklodowskis lived in a small apartment furnished by Mme. Sklodowska's school. After Maria, the youngest, was born, Sklodowski took a teaching post at a boys' high school in Warsaw that provided a larger apartment for the young family. However, as Poland became increasingly Russianized, Wladislaw lost his job, and the family moved to a small house where they took in boarders.

The Sklodowski family stressed the importance of education, and Maria surpassed her siblings, who were also fine students, in academic successes. At kindly Madame Jadwiga Sikorska's excellent private school, as in all Polish schools, the teachers were required to teach only in Russian. However, Maria's

INTRODUCTION

teachers conspired to conceal that they were actually teaching in Polish from Russian inspectors.

Maria's life changed after her sister Zosia died of typhus in 1876 and her mother of tuberculosis in 1878. After these deaths, she rejected the religious beliefs of her childhood and became deeply involved in Poland's problems. Her educational trajectory also changed. Her father moved her from Madam Sikorska's school to the Russian-dominated government-run Advanced High School (Gymnasium) Number Three in downtown Warsaw. This school, previously German, exposed her to a rigorous education. Even though she received an excellent education, especially in physics, Russian literature, and the German language, she at first despised the school. Her popularity among her teachers was damaged when Maria and her friend Kazia were discovered dancing joyously among the desks after the assassination of the Russian tsar, Alexander II. In spite of the problems, she enjoyed school even though she hesitated to admit it. Her life was a roller coaster as she vacillated from joy to deep depression. Like her older siblings Józef and Bronia, she finished first in her class and was awarded a gold medal. Exhausted by the strain of academic achievement and at her father's urging, she took a year off at her uncle's country estate, where she was without the pressure of responsibilities.

When Maria returned to Warsaw, she joined a group of young intellectuals who met to discuss the ideas of positivist philosopher Auguste Comte (1798–1857) and others who advocated social reform. This group was known as a "floating university" and consisted in large part of girls. At this time, Polish girls had only limited opportunities for higher education because they were unable to meet the entrance requirements to the universities in the Russian empire. These institutions required classical languages, which were not taught in the girls' gymnasium. The solution would have been for Maria to attend a university outside of Poland, but the family was unable to afford it. The alternative was to prepare herself for a teaching career in a girls' school by educating herself. Maria developed a plan whereby she

Sixteen-year-old Marie Curie

would first help her older sister, Bronia, attend medical school in Paris by working as a governess, living austerely, and saving her money. After Bronia finished medical school, she would then help Maria with her education.

Maria's first position as a governess in Warsaw was an abject failure, but her second one, although it involved leaving her home city, was more congenial at first. During her three-year tenure (1886–1889) with the Zorawski family, she not only worked seven hours a day as governess to two of their children and the son of a servant, she also taught young peasant children to read and write, with the Zorawski's blessing. In spite of this heavy load, she found time to improve her own knowledge. She educated herself in matters of social change and prepared herself for her future by studying mathematics and physics and reading widely. Toward the end of her tenure, she became increasingly despondent and prone to illness.

INTRODUCTION

She no longer spoke of the Zorawski's as "excellent people," as she had first characterized them. The final affront involved a brief romance with Kazimierz, a son of her employers. The family disapproved of their son's relationship with an employee, whom they considered inferior because of her status.

After leaving the Zorawski's and returning to Warsaw, where she held another position as a tutor, Maria was able to escape after her father accepted the directorship of a reformatory. He was able to send money to Bronia, who was then a medical student in Paris. During this time, Bronia married and invited her sister to come to Paris and share her home while she attended the university. After hesitating for over a year, Maria became a student at the Faculty of Sciences of the Sorbonne in 1891.

Maria found riding to Paris on a fourth-class carriage sitting on a folding chair surrounded by her luggage an unpleasant initiation to her new life. Her brother-in-law, Kazimierz Dluski, met her and took her to their apartment, which she was to share with the couple. This arrangement proved unsatisfactory to Maria (now known as Marie). She left her sister and brother-in-law's apartment for a more convenient but monastically uncomfortable garret in the Latin Quarter. The story of her travails, including severe cold and insufficient food, is well known and described by her daughter Eve in her biography of Marie. Since her allowance from Poland was small and was divided between tuition and the cost of lodging, little was left for necessities such as food. As recorded by her biographers Eve Curie, Robert Reid, and Susan Quinn, she almost starved herself at one point until she was rescued by Bronia's husband, Kazimierz, who brought her back to their apartment where they nursed her back to health. As soon as she got strong enough, she insisted on returning to her attic room and, as her daughter Eve later wrote, "began to live on air."

Marie was at an academic disadvantage as she entered the physical science courses. Although she had to work especially hard to counter her deficiencies, particularly in mathematics, she was fascinated by what she was learning. She wrote in her autobiographical notes, "I divided my time between courses, experimental work, and study in the library. In the evening I worked in my room, sometimes very late into the night. All that I saw and learned that was new delighted me. It was like a new world opened to me, the world of science, which I was at last permitted to know in all liberty." Maria received her degree in physics from the Sorbonne in 1893, after which she returned to Warsaw for a vacation. However, she was not content with her physics degree, for she realized that mathematics was important in developing a deeper understanding of physics, so she returned to Paris to work on a degree in mathematics. Her financial situation was much better as she pursued her mathematics degree because in Warsaw she had been awarded the Alexandrovitch Scholarship for outstanding Polish students who wished to study abroad.

While she was working on her mathematics degree, Marie obtained a job comparing the magnetic properties of different steels. This position involved finding laboratory space for experiments. Hearing of her need, Jozef Kowalski and his wife, whom she had met during her time as a governess, suggested that she meet Pierre Curie, who was working on magnetism at a nearby institution and might have unused space in his laboratory. Although the couple who introduced them may have had matchmaking in mind, any thought of romance was absent from either Marie's or Pierre's mind. Each had experienced unfortunate romantic involvements and vowed to never become involved again. Pierre declared that he would live like a monk, and Marie expected to return to Poland, use her new skills as a teacher, and become involved in Polish politics. Marie had little time to consider romantic entanglements in Paris although she was pursued by an ardent admirer, a Monsieur Lamotte. After her disastrous experience with Kazimierz Zowaski, she was determined never to marry, and when Monsieur Lamotte got serious, she broke up with him.

Fortunately for both and for the world, the couple renounced their previous declarations and embarked upon one of the most fruitful collaborations in the history of science.

INTRODUCTION

Although many 20th-century historians tended to credit Pierre with the theoretical advances made by the pair, recent scholarship has made it clear that their relationship was complementary. As they realized that they shared similar interests and values, they changed their mind about marriage. As Marie later wrote in her biography of Pierre,

> After my return from my vacation our friendship grew more and more precious to us; each realized that he or she could find no better life companion. We decided, therefore, to marry, and the ceremony took place in July, 1895. In conformity with our mutual wish it was the simplest service possible—a civil ceremony for Pierre professed no religion, and I myself did not practice any.

The similarities and differences between Pierre and Marie made for a symbiotic relationship in both their science and their lives that proved advantageous for both of them. Collaboration was familiar to Pierre; for most of his scientific life, he had worked successfully with his older brother, Jacques, investigating properties of crystals. In this relationship, Pierre was the dreamer and thinker, whereas Jacques was the person who translated the thoughts into action. A similar pattern may be detected in the collaboration between Marie and Pierre. Marie might be categorized as a "thinker-doer" and Pierre as a "thinker-dreamer" (Pycior in *Creative Couples in the Sciences*, 46). Several years after their wedding, Pierre willingly put aside his research on crystal growth and worked with Marie on her doctoral research on radioactivity. He contributed to the project by providing a broad knowledge of physics and great skill at designing appropriate instruments. He also made Marie's work more acceptable to male scientists, who would tend to ignore the work of a woman.

As difficult as it was for a woman to be a scientist, it was even more unimaginable for a woman to be both a scientist and a mother. The Curies' first child, Irène, was born on 12 September 1897, after the couple returned from a long bicycle trip. The death of Pierre's

Pierre and Marie collaborating on their work

mother shortly after Irène was born made it possible for Marie to continue her research, for her father-in-law, Èugene, moved in with the couple and gladly took over the care of the baby. In order to have time for her scientific studies, Marie pared down her housework to the bare minimum.

As both Pierre and Marie worked on the research that led to her doctoral dissertation, it evolved from a merely descriptive project to a theoretical one in which she speculated on a cause for the radiation produced by uranium. This project resulted in a major theoretical breakthrough. "My determinations showed that the emission of the rays is an atomic property of the uranium, whatever the physical or chemical condition of the salt were. Any substance containing uranium is as much active in emitting rays, as it contains more of this element." She had demonstrated that radiation was not an ordinary chemical reaction between molecules but issued from the atom itself. Radiation was an atomic property, proportional to the amount of radioactive substance being measured.

The uranium she tested came from the uranium-rich mineral compound known as pitchblende, mined from the Joachimsthal region on the German–Czech border. She found that it produced a current much stronger than that produced by powdered uranium alone. She postulated that the pitchblende must have contained another substance that was more radioactive than uranium. She worked hard in order to isolate this unknown substance from the pitchblende matrix. Through a tedious technique known as fractional crystallization, she was able to isolate two new radioactive elements, polonium and radium, both more radioactive than uranium.

Pierre had never had the academic position that both he and Marie thought he deserved. When the chair of physical chemistry opened at the Sorbonne in 1898, Pierre applied and thought that he would be appointed. However, he had to settle for a less prestigious position teaching physics to medical students. At the same time, Marie got a new paid position at the Normal School (École Normale Supérieure) for girls at Sèvres. To make himself more desirable for a professorship should one open, Pierre agreed to become a candidate for the prestigious Academy of Science (Académie des Sciences) but again was disappointed.

In 1903, Marie defended her dissertation, and the couple was invited to speak at the Royal Institution in London. By this time, both Curies, but especially Pierre, were having health problems caused from their contact with radioactive materials, although they were unaware of the problem. Marie suffered a miscarriage in the same year but again did not blame their research. In December 1903, the Curies and Henri Becquerel were jointly awarded the Nobel Prize for Physics. Their prize was "for their joint researches on the radiation phenomenon discovered by Professor Henri Becquerel" (McGrayne, *Nobel Prize Women in Science*, 25).

Another positive event occurred in 1905, the birth of a healthy baby girl, Eve Denise, on

Portrait of a young Marie Curie soon after the completion of her dissertation

6 December. This birth left Marie exhausted, a foreshadowing of her future health problems. After Eve's birth, Marie returned to her teaching job at Sèvres, although the prize money made it unnecessary. Pierre finally was successful in his second attempt for the French Academy of Sciences in a narrow win. However, his sickness was decidedly worse by 1904. On Thursday, 19 April 1906, Pierre was killed in an accident walking in traffic on a rainy day. He stepped in the path of a horse pulling a wagon, and in his attempt to hold onto the horse slipped, and his head was crushed by the back wheel of the wagon. Marie was distraught at the news of Pierre's death. In an entirely unique move, the council of the Faculty of Science appointed her as an assistant professor. She officially was named Pierre's successor in the chair that he had only occupied for 18 months. She was the first woman to have such a position at the Sorbonne.

In her new position, Marie made both friends and enemies. She had been the subject of both positive and negative publicity as she unsuccessfully sought membership in the academy. As far as anyone knew, Marie remained a grieving widow without a romantic interest. However, this was soon to change when a thief broke into Paul Langevin's study and stole letters from Curie that implied a close relationship between the two. Langevin had been Pierre Curie's Ph.D. student and was a respected scientist, a husband, and the father of four children. The scandal broke, and Curie was vilified in the newspapers and her reputation impugned by scientists and citizens alike, as she was characterized as a home wrecker while her scientific achievements were ignored. Five duels were provoked by the Langevin/Curie affair. In the midst of the scandal, she went to Stockholm to accept the Nobel Prize for Chemistry, for isolating radium. Although she managed to get through her acceptance speech, the repercussions from the affair were severe. She had numerous physical problems and suffered from severe depression, even going so far as to use her maiden name, Sklodowska, because she did not want to besmirch the name Curie. However, by December 1912, she began her experimental work again using the name Curie. The problems with the affair had receded into the background.

Radium institutes had become common in Europe with the main emphasis on radium as a treatment for cancer. Curie was able to convince the Pasteur Institute and the Sorbonne to establish an institute devoted to the science of radioactivity. However, Curie's institute was completed just as Germany declared war on France. Marie sent her daughters to the countryside and took the radium that was still in her laboratory to Bordeaux for safe keeping. During the war, Marie organized radiology services for military purposes.

The most creative period scientifically for Curie occurred in the years before the war. Even though she returned to the laboratory after the war, much of her later work involved raising funds by virtue of her reputation to fund radiological research at her new institute. After the war, France was impoverished and thus unable to purchase the radium that was needed for research on the institute's major emphasis, radiation as a treatment for cancer. Also, Curie's reputation in France was still tarnished because of the Langevin affair. The United States, relatively unscathed by the war, represented a possible source for the expensive substance. Marie Mattingly Meloney (Missy), the editor of an American women's magazine, *The Delineator*, campaigned in the United States to collect money to purchase a gram of the precious radium so that cancer research could continue. Although Curie was not active in the fund raising, she wrote that if Missy was successful in raising the money, she would, in spite of her reservations, come to the United States. Shy Marie who despised crowds was overwhelmed by the exuberant American reception. After accepting the radium from President Warren Gamaliel Harding, Curie gladly would have canceled the rest of her trip, but that was not to be.

Not always happily, she accepted honorary degrees while suffering from exhaustion and illness. Although she was scheduled for a trip to the West involving many social activities, the social events were canceled, and she and Meloney tried, unsuccessfully, to make the

INTRODUCTION

Marie Curie and President Warren Harding during Curie's visit to the United States in 1921

western trip less strenuous. On her return to France, Marie noted that her work had been made easier by the gift of radium. This success inspired her to work harder to obtain more funding for her institute.

The idea of a laboratory dedicated to research on radioactivity was originally Pierre's. Marie carried on his hope that such an institute would materialize, especially after a chair at the Sorbonne had been instituted in his honor. The physician director of the Pasteur Institute, Émile Roux, was an important advocate for this institution. He proposed that his well-funded institute could build the laboratory. The university agreed and with Roux' help planned two separate laboratories, one directed by Curie and funded by the university, and the second studying the medical applications of radioactivity and directed by Dr. Claudius Regaud. Together, these two buildings would be known as the Institut du Radium.

By the late 1920s, both of the two original laboratories had expanded and the personnel grown. Curie herself was involved in the supervisory aspects of the laboratory but was less involved in the actual research. This laboratory was diverse, with many women involved, many of whom developed into important scientists themselves. In addition to the women, the lab was ethnically and nationally diverse.

During this time, Curie's health continued to degenerate, but she was loath to admit that her child, radium, had anything to do with it. However, as radium developed into an industry and more reports came in, it became more and more difficult to ignore its harmful effects. She was forced to admit that certain local effects of radium could cause harm, such as sores on her fingers where she handled the substance. However, the fact that symptoms experienced by those who came in contact with radioactive materials varied greatly left her unconvinced that radioactivity was the culprit. It seemed inconceivable to her that radiation could both cause and cure cancer. The same reasoning made it difficult to explain how the cataracts she had developed were an early sign of exposure to radiation when certain physicians were using radiation to cure them. When the workers in her laboratory complained about fatigue, she rationalized that it could be caused by any number of circumstances.

As reports from radium laboratories and factories increased, Curie still was unable to admit to her own laboratory workers that radium was dangerous. She did, however, acknowledge to a Polish laboratory worker, Alicja Dorabialska, that she did not fully understand radium's effect on health.

Marie Curie's two daughters were very different. Irène had long been Marie's favorite child, probably because their interests and personalities were similar. Eve found the arts and humanities more interesting than the sciences. However, after Marie was homebound following cataract surgeries, it was Eve who became her domestic main support. Irène was her companion in scientific enterprises and worked with her mother in the laboratory. She announced her engagement to a fellow laboratory scientist, Frédéric Joliot, to her mother's surprise and shock. Marie's fear that she would share her daughter with someone else was frightening to her. Marie reluctantly accepted the marriage, and Frédéric later decided to hyphenate their names and become the Joliot-Curies. The couple won the coveted Nobel Prize in 1935 for their work in artificial radioactivity.

In her later years, Marie Curie spent more time raising money for her laboratory and less in creative scientific activities. She also felt a great responsibility to her birth home, Poland, and sought the help of Missy Meloney to raise money for an Institute of Radium in Warsaw. They raised the money just before the stock market crash on 24 October 1929. Marie made a second trip to the United States to thank the citizens for their help.

Although her health was increasingly fragile, Curie refused to retire. However, she was less active in the laboratory than before and her output of new scientific papers decreased. Some of her colleagues complained that she was growing increasingly testy, but the young scientists who came to the laboratory that she mentored were very loyal to her.

She was also helpful to the scientists whom she considered worthy. Nevertheless, she relinquished many of her laboratory responsibilities as director to Irène and Frédéric, causing some dissension among the workers. As her health declined, family became more important to her, and she especially enjoyed her grandchildren, Irène and Frédéric's children, Hélène and Pierre.

During the winter of 1934, when her health was obviously miserable, she still went ice-skating and skiing and refused to admit that she was very ill. She must have realized that her death was eminent, for she explained to Irène where all of her personal documents as well as relevant scientific ones were located. Misdiagnosed with tuberculosis, she was admitted to a sanatorium after becoming acutely ill on a trip to the south of France with two of her sisters. She died on 4 July 1934 and was buried in the same grave where Pierre had been placed. The remains of both Pierre and Marie were transferred from the tomb in Sceaux to the Pantheon in Paris on 20 April 1995.

ACADÉMIE DES SCIENCES (ACADÉMIE ROYALE DES SCIENCES; FRENCH ACADEMY OF SCIENCES). This prestigious society was founded in 1666 by Louis XIV through the efforts of his minister of finance, Jean-Baptiste Colbert (1619–1683). During the 17th century, academies were founded in several European countries to promote the arts, literature, and science. When the French society was first founded, Colbert invited a small group of scholars versed in physics and mathematics and those from various disciplines such as history and the so-called belles lettres to meet in the king's library. The scholars from the various disciplines divided up, and the geometers and physicists agreed to meet twice weekly (Wednesdays and Saturdays) in a room in the king's library that contained books on physics and mathematics. The groups met together once each month, and their secretaries reported on accomplishments of the preceding month. Unlike its British counterpart, the **Royal Society of London**, the French organization was founded as an organ of government. After the society's champion, Colbert, died, the academy was in decline until it was reorganized in 1699. After the French Revolution (22 August 1795), the National Institute of Sciences and Arts replaced the earlier institution, and the members of the old academy were formally reelected. Membership was not restricted to scientists. For example, Napoleon Bonaparte was elected a member. The society's name returned again to the Académie des Sciences during the Second Republic. The academy's proceedings were published under the name of *Comptes rendus de l'Académie des sciences* (1835–1965); the *Comptes rendus* is now a journal series. In the 1920s, the center of power shifted from the older conservative Académie. In spite of the Curies' difficulties with the Académie, **Pierre Curie** finally was admitted, although **Marie Curie** never was. In spite of their difficulties, the Académie provided early financial support for the Curies. Marie was a three-time recipient of the Prix Gegner (**Gegner Prize**), and Pierre won the 10,000-franc biannual **Prix La Caze**. They also benefited from the institute's 20,000-franc credit available to them for their work in isolating radium. Although Marie never became a member, 51 years later one of her researchers, **Marguerite Perey**, became the first woman elected to the Académie des Sciences. *See also* POSITIONS, PROBLEMS WITH, FOR P. CURIE.

ALPHA PARTICLES. The existence of these particles, consisting of two protons and two neutrons bound together into a particle identical to a helium-4 nucleus, was postulated by **Ernest Rutherford**, a student of **Joseph John (usually referred to as J. J.) Thomson**, in his search for a theory of matter. Thomson had already taken a step in cutting the atom when he proposed the existence of electrons and postulated that they were the source of electricity. Although Rutherford was not especially interested in the Curies' discovery of new elements, he recognized in them the possibility of a further "cut" in the atom. It was Rutherford who explained the phenomenon of **radioactivity** by proposing that radioactive emissions were composed of at least two different "rays," beta rays (identical to J. J. Thomson's negatively charged particles) and alpha rays, positively charged particles. It was many years later that the actual morphology of the atom was better understood. Rutherford's explanation seemed to answer the original conundrum of how radioactivity was related to atomic makeup. His theory involved the behavior of the radioactive element **thorium**, an element that **Marie Curie** and **Gerhard Carl Schmidt** had discovered independently. When he attempted to measure the ionizing power of thorium oxide, Rutherford was intrigued by puzzling data. He found that thorium gave off a gaseous substance, an "emanation" that had the power of producing radioactivity in any substance it came in contact with. The emanation from thorium had properties the thorium itself did not have. Rutherford and his student **Frederick Soddy** demonstrated that this emanation temporarily carried off thorium's radioactivity. **Pierre Curie** and **André-Louis Debierne** found that the same phenomenon occurred with radium. Rutherford proposed a hypothesis of transmutation. The radioactive substance is an atomic phenomenon with an accompanying chemical change in which new kinds of matter are produced. The Curies were suspicious of this transmutation hypothesis, possibly because it might detract from their discovery of polonium and radium if these elements were so unstable as to transmute into different materials.

AMAGAT, ÉMILE (1841–1915). Amagat was the French physicist who competed with **Pierre Curie** for membership in the **Académie des Sciences**. He won the election on 9 June 1902 with 32 votes to Pierre's 20. Pierre was a reluctant candidate for the position, but once he was convinced by his fellow physicists to compete for the honor, he was disappointed with the result of the election. Amagat had made a great effort to convince the voters that his seniority (he was 18 years older than Pierre) meant that he deserved the honor. His work on the isotherms for different gases, his Law of Partial Volumes (1880), and his invention of the hydraulic manometer provided the basis for his election.

ANARCHIST MOVEMENT. The historic Anarchist Movement was a workers' movement flourishing from the 1860s through the 1930s in the West, although it had its origin much earlier. Its manifestations differed in various parts of the world. From the Greek root *anarchos*, anarchist thought focused on the idea that government is both harmful and unnecessary. During the Third Republic in the 1890s, Paris was schizophrenic. Although it was innovative in culture and technological developments, it was beset by a wave of anarchy and a new café society that many considered decadent. The Industrial Revolution with its advances in technology produced the tallest structure in the world, the Eiffel Tower. Electric lights, moving pictures, and electric streetcars were exciting, but rather than improving life, they resulted in increased poverty, dirtier cities, and a frantic lifestyle. The French government was corrupt and mired in anti-Semitism and infested with scandal, including the Dreyfus affair, where an army trainee by the name of **Alfred Dreyfus** was falsely accused of treason in a wave of anti-Semitism. In response to governmental chaos, the international anarchist movement came to Paris, where it flourished. From 1892 to 1894, 11 anarchist bombs exploded in Paris, finally resulting in the assassination of the president of the republic, Sadi Carnot, by an Italian anarchist. Some members of the artistic community considered the anarchists martyrs and blamed members of the positivist scientific community for the chaos. People tended to think that science with its emphasis on reason and apparent worship of **positivism** was at least partially responsible for the turmoil. Some of the most exciting scientific breakthroughs emerged in this period, including those of **Marie** and **Pierre Curie**, who thought that salvation lay in science and reason.

ANODE. An anode is a positively charged electrode. *See also* RÖNTGEN, WILHELM.

APPELL, PAUL (1855–1930). Appell was **Pierre Curie**'s senior colleague and dean of the faculty of science at the **Sorbonne**. He brought the news of Pierre's death first to Pierre's father and then to Marie when she returned home. *See also* DEATH OF P. CURIE.

AYRTON, HERTHA MARKS (1854–1923). Born Phoebe Sarah Marks, Ayrton was the third child of a Polish Jewish refugee, Levi Marks, who had difficulty in providing for his wife and eight children with his clock-making/jewelry trade in Portsea, England. After his death in 1861, his wife, Alice Theresa, tried to support the children with her needlework. As a child, staunchly independent Phoebe helped care for her younger siblings and learned homemaking skills from her mother. She was only able to attend school because she had an aunt who ran a school in London. Unconventional Phoebe rejected her real name and adopted a new one, Hertha. An eccentric philanthropist, **Barbara Bodichon**, interested in **women**'s causes and one of the founders of **Girton College, Cambridge**, befriended Hertha and made it possible for her to attend the new institution. After leaving Girton, she first became a mathematics teacher but then resigned to tutor private students. During her teaching career, she met her future husband, the much older Professor **W. E. Ayrton**, a fellow of the **Royal Society**, advocate for technical education, and supporter of women's rights. The independent personalities of both Hertha and W. E. made their relationship more a mutually supportive one rather than the true collaboration of the Curies. However, the women in both cases found success in science much more likely because of their scientist husbands. Hertha became known for her contributions to studies of the electric arc. She had some experiences that were similar to those of **Marie Curie** in that she was denied membership in the Royal Society, as was Curie in the **Académie des Sciences**. They also shared the experience of children. W. E. Ayrton had a daughter, Edith, and they had one daughter together, Barbara Bodichon Ayrton, and, of course, the Curies had two daughters, Irène and Eve. They also shared a passionate involvement in social justice issues, Curie especially in issues having to do with Polish independence and Ayrton as a supporter of women's causes. Curie, however, as she became more involved in science, gave up political activism for pure science, whereas Ayrton supported the independence of Ireland and was very active in the English suffrage movement. The two women met in June 1903, when Pierre addressed the Royal Society on the discovery of **radium**. As the wife of physicist W. E. Ayrton, Hertha was present in that role, not as a scientist in her own right. The Curies were invited to the Ayrtons during that visit, and Marie and Hertha developed a friendship that was lifelong and corresponded frequently. After Pierre's death, Ayrton invited Marie Curie and her daughters to join her in England at the seashore. After the Langevin scandal, she finally agreed. Traveling under the name "**Madame Sklodowska**," she was suffering both physically and psychologically. Hertha was able to keep Marie's identity safe from the press, and although she was still in pain the trip was therapeutic. In October 1912, Marie was strong enough to take the ferry from Dover to Calais and then to Paris, where she quickly rejoined the scientific scene. *See also* CURIE–LANGEVIN AFFAIR.

AYRTON, W. E. (1847–1908). W. E. Ayrton was the electro-physicist husband of **Hertha Ayrton**. Ayrton was a widower with one daughter, Edith. He believed in equality of opportunity between men and **women**. After W. E. became ill in 1905, he relied on his wife to complete his scientific commissions. Proud of his wife, he insisted that she get the credit that she deserved. On the fourth report, he insisted that Hertha's name be included as the author of the publication. He also was very supportive when Hertha became more involved in the militant wing of the suffrage movement.

B

BECQUEREL, ALEXANDRE-EDMOND (1820–1891). Alexandre Becquerel was the second of three eminent Becquerel physicists. Antoine-César Becquerel (1788–1878) was a professor at the Paris Museum of Natural History, and his son was **Antoine-Henri Becquerel (1852–1908)** shared the Nobel Prize in 1903 with the Curies. Alexandre-Edmond invented an instrument to identify new substances with phosphorescent qualities even when they only phosphoresced for a brief time. His son, Antoine-Henri, took up his father's interest and became fascinated by **phosphorescence**.

BECQUEREL, ANTOINE-HENRI (1852–1908). Antoine-Henri (known as Henri) Becquerel, the son of **Alexandre-Edmond Becquerel**, became involved in his father's interest, **phosphorescence**, after he heard about **Wilhelm Röntgen**'s discovery of X-rays and **Henri Poincaré**'s discovery that some substances continued to glow without the presence of light. Although Henri Becquerel held a doctorate from the Sorbonne, he was not active in research until he heard Poincaré's report on X-rays. Henri Becquerel and three other scientists hypothesized that the phosphorescent substance itself could produce X-rays and that a **cathode** ray tube was unnecessary. The other three scientists were convinced that this hypothesis was correct, but Henri Becquerel did not find X-rays when he experimented on phosphorescent substances. However, when he exposed a sample of **uranium** salts, he found that they immediately produced radiation. In a report to the French Academy, he explained his methodology. He wrapped a photographic plate in two sheets of thick black paper to protect the plate from sunlight. Then he placed a plate of a phosphorescent substance above the paper and exposed the entire package to the sun for several hours. When he developed the photographic plate, he found that the silhouette of the phosphorescent substance appeared in black on the negative. He then placed a coin between the phosphorescent material and the paper and exposed it to the sun. He found that the coin's image appeared on the negative. His conclusion was that the phosphorescent substance that he used emitted rays that could penetrate paper impervious to light. He concluded that it was the sun that allowed the material to phosphoresce and penetrate the photographic plate. To corroborate this result, he prepared another experiment. He placed a copper cross between the black paper covering the plate and the uranium salts. He assumed that when the package was exposed to the sun, the pattern of a cross would appear on the plate. But it was February in Paris, and the sun did not shine. Becquerel put the entire experimental setup in a dark cabinet until the weather improved. Becquerel became impatient and developed the plate even without its exposure to the sun. He had assumed that the plate would be blank because it had not been exposed to sunlight. However, the pattern of the cross appeared. He concluded that it was the uranium in the mixture that caused the image to appear on the plate. Nevertheless, he continued to believe that phosphorescence was involved in some way in the phenomenon. He thought that a form of phosphorescence was stored in the uranium and that its emission was a kind of invisible phosphorescence. Although he had actually discovered **radioactivity**, he did not name it nor explain its source. The explanation for the phenomenon was left for **Marie Curie** to discover. See also NOBEL PRIZE FOR PHYSICS, 1903, M. AND P. CURIE; H. BECQUEREL; *RESEARCHES ON RADIOACTIVE SUBSTANCES.*

BERZELIUS, JÖNS JACOB (1779–1848). Swedish chemist Berzelius discovered the element **thorium** in 1828. When **Marie Curie** measured the activity of other elements against that of pure **uranium** and **pitchblende**, she found that the mineral aeschynite, which contains thorium but not uranium, was also more active than uranium. She found that pitchblende was also more active than uranium. These results led her to recognize that "activity" was not a unique characteristic of uranium but a more general phenomenon.

BICYCYCLING VACATIONS. In the early years of their marriage, **Marie** and **Pierre Curie** often went on bicycling vacations in the countryside near Paris or sometimes to the mountains along the sea. They only left Paris for short periods of time because they hesitated to leave their work. They cut short a cycling trip to Brest when Marie was eight-months pregnant and returned to Paris, where Irène was born on 12 September 1897. Throughout their lives together, bicycling was a favorite activity. See also BIRTH AND INFANCY OF I. JOLIOT-CURIE.

BIRTH AND INFANCY OF I. JOLIOT-CURIE. Irène was the older daughter of **Marie** and **Pierre Curie**. She was born after one of Marie and Pierre's **bicycling trips**. In September 1897, Pierre's mother was terminally ill with breast cancer, and Marie suggested that a cycling trip would be therapeutic for Pierre in spite of the fact that she was eight-months pregnant. Neither seemed to be concerned about the imminent birth. However, they returned to Paris early. Pierre's mother died two weeks before the baby was born. Shortly after their return from the cycling trip, Marie gave birth to a 6.6-pound baby girl, Irène. Marie never considered giving up her research. She was convinced that she could handle both her career and motherhood. Shortly after Irène was born, Marie began a notebook in which she recorded the important events in her children's lives for the next 15 years. Although Marie attempted to nurse Irène, the baby did not thrive, so they were obliged to obtain a wet nurse for her. Marie was a very solicitous mother and, although they had both a baby nurse and a wet nurse, took care of most of the baby's needs, such a changing, bathing, and dressing her. The nurse took Irène to the laboratory, where her mother was finishing and editing her work on magnetization. See also PREGNANCIES OF M. CURIE.

BLACK LIGHT. After **Wilhelm Röntgen** discovered X-rays, others claimed to have found other different forms of radiation, several of which were hoaxes or, at the very least, misunderstandings. The name **Gustave LeBon** gave to one of these "new" forms of radiation was "black light." See also BLONDLOT, RENÈ; N RAYS.

BLONDLOT, RENÉ (1849–1930). Blondlot claimed to have discovered new rays that he called **N rays** after his native city of Nancy. He claimed to have produced these rays by placing a hot wire inside an iron tube. They could be detected by a calcium sulfide thread that glowed slightly in the dark when the rays were refracted through a prism and produced a spectrum. **Gustave LeBon** claimed that the N rays were invisible except when they encountered the thread. After he reported his findings, laboratories all over the world claimed to have generated N rays. See also BLACK LIGHT.

BODICHON, BARBARA (1827–1891). As a British education and political activist for the rights of **women** and one of the founders of the women's college **Girton College, Cambridge**, Bodichon helped numerous women get an advanced education. Marie Curie's friend **Hertha Ayrton** was one who benefited from her help.

BOGUSKA, BRONISLAWA. *See* SKLODOWSKA, BRONISLAWA.

BOGUSKI, HENRYK (1830–1913). Henryk Boguski was Maria Sklodowska's (**Marie Curie**'s) uncle who, along with his brother Wladyslaw, was considered a black sheep of his family. Henryk lived on his wife's income as manager of the village store and became involved with his brother in failed moneymaking schemes. After Maria graduated from **Gymnasium Number Three** in 1883, where she finished first in her class and was awarded a gold medal, she was stressed and tense. The solution seemed to be a relaxing time with her mother's brothers in the country. She was unaware of the financial difficulties of her uncles and enjoyed the carefree atmosphere of their household, where she had little responsibility to work or study. She slept late, read novels, walked in the woods, and generally had a good time, forgetting for a while the pressure she felt to excel in school and the sadness that enveloped her on the death of her mother (tuberculosis) and sister Zofia (Zosia) (typhus). *See also* BOGUSKI, JÓZEF JERZY.

BOGUSKI, JÓZEF JERZY (1853–1933). While still in **Warsaw**, Marie Sklodowska (Curie) in her self-educating phase taught herself chemistry without a laboratory. She found this frustrating and wrote her chemist cousin, Józef Boguski, 14 years older than Sklodowska, asking him to provide her with the use of a small laboratory with apparatus for simple experiments in chemistry and physics. Boguski had served as an assistant to the Russian chemist Dmitri Mendeleyev and returned to Warsaw to run a laboratory at what he called the Museum of Industry and Agriculture. He apparently chose this grandiose name to hide the fact that it was one of the clandestine teaching organizations in Warsaw to hide from the Russians. Sklodowska tried different experiments described in the textbooks, and this was possibly the reason she chose to study chemistry and physics rather than biology. Later, Boguski became a professor at the Wawelberg and Rotwand School and from 1920 was a professor at the Warsaw Polytechnic University. He became known for his pioneering studies in chemical kinetics.

BOLTWOOD, BERTRAM BORDEN (1870–1927). Marie Curie made an enemy of the American radiochemist Bertram Boltwood. He was a Yale professor for three years (1897–1900) and established lead as the final product of the decay of **uranium**. He suffered from depression and committed suicide on 15 August 1927. His problem with Curie surfaced when he asked her to allow him to compare one of his **radium** solutions with her radium standard. When she refused, Boltwood accused her of being unwilling to help any worker in **radioactivity** outside of her own laboratory. **Ernest Rutherford** finally convinced Curie to lend him the standard. *See also* RADIUM, INTERNATIONAL STANDARD FOR.

BOREL, ÉMILE (1871–1956). Mathematician Émile Borel and his wife Marguerite were social friends of the Curies. *See also* BOREL, MARGUERITE.

BOREL, MARGUERITE (1883–1969). Marguerite Borel was the much younger wife of mathematician Émile Borel. She was a friendly person who invited confidences from people with problems. She had a role in the **Paul Langevin** scandal five years after **Pierre Curie** died. She reported that suddenly Marie gave up her somber dark clothes and appeared at the home of physicist **Jean Perrin** and his wife in a white dress with a rose at her waist. Since Paul Langevin had confided to her about his unhappy marriage and Marie had also excoriated Langevin's wife for not supporting him, she drew the conclusion that the two were involved romantically. *See also* CURIE–LANGEVIN AFFAIR.

BOUSSINESQ, JOSEPH VALENTIN (1842–1929). Boussinesq was a French mathematician and physicist who was one of **Marie Curie**'s teachers at the **Sorbonne**. Although he was opposed to **Albert Einstein**'s ideas on relativity, he was an excellent physicist of the old school and taught her the details of classical physics. *See also* SORBONNE, M. CURIE AS A STUDENT AT THE.

BRANLEY, EDOUARD (1844–1940). Branley was **Marie Curie**'s opponent for membership in the **Académie des Sciences**. He had been a candidate for the award two times previously and had been defeated both times. His contribution was in technology. He made "radio conductors" by filling tubes with iron filings that could receive electromagnetic signals. Many claimed that he should have received a Nobel Prize in Physics along with **Guglielmo Marconi** (1874–1937). The right-wing press was pleased to find a male candidate to oppose Curie. The election pitted the liberals, feminists, and anticlerics, Curie's supporters, against the nationalist, pro-Catholic, anti-Semitic right. The actual vote, on 23 January 1911, was 28 for Curie, 30 for Branley, and one for a third candidate. Curie was extremely disappointed and never stood for membership in the academy again.

BUNSEN, ROBERT (1811–1899). The technique of spectroscopy used to identify previously undescribed elements was discovered by Robert Bunsen and **G. R. Kirchoff** in the 1860s. The Curies used this technique to test **pitchblende** for the substance that was making it more active. However, it did not produce the desired results.

CARLOUGH, MARGUERITE (1901–1926). Carlough was one of the "**radium** girls" who became ill after painting the dials of clocks so that they would glow in the dark. In order to make the numbers more accurate, they sharpened their brushes by touching them to their lips and tongues. She was the first to bring a lawsuit against dial-painting firms. The publicity surrounding the lawsuit led to the Waterbury Clock Company banning "lip pointing." *See also* RADIUM, HEALTH PROBLEMS FROM.

CARNEGIE, ANDREW (1835–1919). Marie Curie met Scottish American industrialist and philanthropist Andrew Carnegie in Paris shortly after **Pierre Curie**'s death. He was impressed by Curie and endowed her research. The endowment made it possible for her to pay a research staff around which she could build an institute of **radioactivity** in Paris. Many years later, during Curie's visit to the United States, Mrs. Andrew Carnegie sent her car to pick up Marie and take her to the Carnegie mansion, where she was probably entertained. *See also* UNITED STATES, M. CURIE'S 1921 TRIP TO THE.

CATHODE. A cathode is a negatively charged electrode. **Wilhelm Röntgen** investigated the properties of cathode rays, or negatively charged particles (**electrons**), emitted by a high-vacuum discharge tube. He found that when activated by a high-voltage current, electrons would race from the cathode to the **anode** (positively charged electrode). The phenomena that resulted from this led to experiments that led to the discovery of X-rays.

CHAVANNES, ALICE (1868–1927). To ensure that Irène and Eve had the kind of education she desired, **Marie Curie** organized a cooperative school for approximately 10 children with the parents of the children and friends serving as teachers. Alice Chavannes, whose daughter Isabelle was Irène's friend, taught English, German, and geography. *See also* CHILDHOOD AND EDUCATION OF I. JOLIOT-CURIE.

CHILDHOOD AND EDUCATION OF I. JOLIOT-CURIE. Marie Curie was involved in the education of both of her children. She believed that it was important for them to participate in sports and spend time away from their books in the open air. Although Irène had a year of traditional education beginning when she was 10 years old, Marie wanted her to have a more challenging educational environment. In order to ensure that Irène got the kind of education that she approved of, Marie organized a cooperative school with approximately 10 children attending. The faculty included the parents of the students and talented friends who were willing to help. This plan resulted in a school with some of the most brilliant individuals serving as teachers in France. The students met in various homes but sometimes visited museums in Paris and other nonacademic places rather than having a formal class. The curriculum was varied and included not only science and mathematics but also sculpture and Chinese. Marie Curie taught chemistry; **Jean Perrin**, physics; **Henriette Perrin**, history and French; **Paul Langevin**, mathematics; and **Alice Chavannes**, English. The students had plenty of time for extracurricular activities such as gymnastics, swimming, bicycling, rowing, skating, and horseback riding. This school only lasted for two years because the parent teachers were too busy to continue the project. Irène then entered a more traditional learning environment at the Collège Sévigné in central Paris from 1912 to 1914, completing her baccalaureate degree in 1914. Irène then entered the Faculty of Science at the **Sorbonne**, but her studies were interrupted by **World War I**. She resumed them after the war and received degrees in physics and mathematics. She earned her *licence* in 1920 and her D.Sc. in 1925, with a thesis on the **alpha** rays of **polonium**.

COMMUNE OF PARIS OF 1871 (MARCH TO MAY 1871). After France's defeat in the **Franco-Prussian War** and the crushing settlement of 1871, rebels in Paris refused to disarm and submit to the French interim government supported by the Prussians. They formed the Commune of Paris. A bloody battle occurred between those loyal to the new French government and the commune. The result was the suppression of the commune and the eventual establishment of the Third Republic. **Marie** and **Pierre Curie** experienced the result of these events when they were first in Paris.

COMPTES RENDUS. See ACADÉMIE DES SCIENCES.

COMSTOCK, ANNA BOTSFORD (1854–1930). Anna Comstock's life as a naturalist and proponent of the nature study movement exemplifies a strategy used by women in order to gain a foothold in the sciences: collaboration with a male partner. **Marie Curie** found it easier to be accepted by the male scientific community because of her partnership with **Pierre Curie**. As was the case with the Curies, Anna Botsford when she married entomologist **John Henry Comstock** was able to make her own career in natural history education at Cornell University. As a part of the Cornell extension program, she wrote and illustrated a set of *Nature Study Leaflets* and also was active in nature education all over New York.

COMSTOCK, JOHN HENRY (1849–1931). John Henry Comstock was a well-known entomologist and arachnologist at Cornell University. He married Anna Botsford in 1878, and they became examples of scientific collaboration. This partnership, as did that of the Curies, allowed the female partner to achieve more of her potential because she became more acceptable to the scientific community. *See also* COMSTOCK, ANNA BOTSFORD.

COMTE, AUGUSTE (1798–1857). Comte, a 19th-century philosopher, is known as the founder of the positivist movement, a philosophy that claimed sensory experience was the most perfect form of knowledge and, consequently, information derived from sensory experience and interpreted through reason is the exclusive form of all certain knowledge. **Poland** developed its own breed of **positivism** to include Poland's political and socioeconomic problems. As a young student in **Warsaw**, Marie Sklodowska was involved in its study along with her friends.

COOLIDGE, CALVIN (1872–1833). Coolidge, a Republican, was vice president of the United States in 1923 when President **Warren G. Harding** unexpectedly died. "Silent Cal," as he was known, succeeded Harding in the White House. While he was vice president he wrote an article for an American women's magazine, *The Delineator*, edited by the conservative **Marie Meloney**, in which he upbraided women who worked outside the home and left the care of their children to others. Although Meloney agreed with Coolidge's ideas, she still rationalized **Marie Curie**'s absence from her children in an editorial about **radium** in the same magazine that carried Coolidge's rant.

CROOKES, SIR WILLIAM (1832–1919). Crooks was a British chemist and physicist. He is probably best known for inventing a vacuum tube, the Crookes tube, in 1875. He investigated properties of **cathode** rays, showing that they travel in straight lines and when they fall on certain substances they cause fluorescence. He believed he had discovered a fourth state of matter, which he called radiant matter. In the late 1860s, he became interested in spiritualism and studied the work of several mediums between 1871 and 1874, whom he believed could produce paranormal phenomena. Several other turn-of-the-century scientists, notably Alfred Russel Wallace (1823–1913) and Charles Robert Richet (1850–1935), were also fascinated by the idea of a spirit world. Although both **Pierre** and **Marie Curie** attended séances by medium Eusapia Palladino, Pierre especially was interested in the fact that he could not find an obvious way to discredit the spirits they conjured up. Marie did not share Pierre's interest in spiritualism.

CRYSTALLOGRAPHY. Crystallography, the experimental science of studying the arrangement of atoms or molecules in certain mineral substances, was the research subject of the Curie brothers, Jacques and Pierre. They were fascinated by symmetry in nature. As they studied crystals with multiple planes and axes of symmetry, they discovered a new characteristic of a phenomenon previously discovered by Jacques's teacher, that heat caused these crystals to become charged. This phenomenon was called pyroelectricity, the phenomenon of generation of small amounts of electricity from heat. The Curie brothers determined that the cause of the charge was not heat but pressure. When certain of these crystals were altered by pressure, they gave off an electric charge. They also discovered the opposite—when an electric charge was applied to a complex crystal, it would change shape. They published numerous papers on the phenomena, which came to be called **piezoelectricity**. *See also* CURIE, PAUL-JACQUES; RESEARCH OF P. CURIE.

CURIE, EUGÈNE (1827–1910). Eugène Curie, the father of **Pierre** and **Paul-Jacques Curie** (usually referred to as "Jacques") and the father-in-law of **Marie Curie**, was a physician who, in his younger days, was an advocate of revolutionary ideas, including anticlericalism and egalitarianism. Although it is probable that Eugène Curie's radical ideas prevented him from getting well-paid medical positions, he had an important influence over his sons. Neither boy was baptized nor exposed to religion. After his wife's death, Eugène lived with Pierre and Marie. Eugène was alone at the house when **Paul Appell** and **Jean Perrin** arrived at the Curie house to inform Marie of Pierre's death. She had not yet arrived home from work. When he saw the expression on Appell's and Perrin's faces, Eugène knew that his son was dead and burst into tears while accusing Pierre of absentmindedness. After Pierre's death, Eugène became even more an important member of the Curie household. His presence as a grandfather to **Irène** and **Eve Curie** made it possible for Marie to continue with her scientific work without guilt, knowing that the girls were being well cared for. His playful teasing and ability to lighten the atmosphere of the home made life more pleasant for the sisters. He became ill in 1909 and was confined to his bed for a year. Apparently, he was a difficult patient, and Marie spent much of her time trying to distract him from his illness. He died on 25 February 1910. Marie requested that the gravediggers remove Pierre's coffin and place Eugène's at the bottom of the grave with Pierre's coffin above it, so that at her death the empty space above Pierre would be for her. In death, she wanted to be near Pierre. *See also* MOTHER, M. CURIE AS A.

CURIE, EVE DENISE (1904–2007). After **Marie Curie** had a miscarriage in 1903, she became pregnant again, and Eve, a perfect baby girl, was born on 6 December 1904. Since her father died in 1906, Eve did not have memories of Pierre. Eve, though very talented in her own right, was uninterested in science. Her interests were more literary and artistic; she was also a talented musician. *See also* LABOUISSE, EVE DENISE CURIE.

CURIE, IRÈNE. *See* JOLIOT-CURIE, IRÈNE.

CURIE, MARIE (MARIA) (1867–1934). When discussing her life before she married **Pierre**

Curie, she is referred to by her maiden name, Maria Salomea Sklodowska. Maria was born in **Warsaw, Poland**, the youngest of the five children of Bronislawa Boguska, the headmistress of a fine private school for girls in Warsaw, and **Wladyslaw Sklodowski**, a mathematics and physics teacher. See the following entries concerned with different aspects of her life and works: BICYCLING VACATIONS; CURIE–LANGEVIN AFFAIR; DEATH OF P. CURIE; DOCTORAL DISSERTATION OF M. CURIE; DUELS INVOLVING M. CURIE; EARLY RESEARCH OF M. CURIE; GOVERNESS, M. CURIE AS A; HEALTH PROBLEMS OF M. CURIE; "MADAME SKLODOWSKA"; MOTHER, M. CURIE AS A; NOBEL PRIZE FOR CHEMISTRY, 1911, M. CURIE; NOBEL PRIZE FOR PHYSICS, 1903, M. AND P. CURIE AND H. BECQUEREL; PREGNANIES OF M. CURIE; *RESEARCHES ON RADIOACTIVE SUBSTANCES*; SÈVRES, TEACHING POSITION AT; SORBONNE, M. CURIE AS A PROFESSOR AT THE; SORBONNE, M. CURIE AS A STUDENT AT THE; STUDENT, M. CURIE AS A PREUNIVERSITY; UNITED STATES, M. CURIE'S 1921 TRIP TO THE; UNITED STATES, M. CURIE'S 1929 TRIP TO THE; WORLD WAR I, EXPERIENCE OF M. CURIE IN.

CURIE, PAUL-JACQUES (1855–1941). Jacques Curie, the older brother of **Pierre Curie**, was a French physicist who collaborated with his younger brother in their study of pyroelectricity in the 1880s. The brothers began to work together when Pierre was 21 and Jacques 24. Jacques was appointed professor of mineralogy at the University of Montpellier in 1883, which ended the brothers' collaboration. Jacques remained at Montpellier, where he was appointed to the chair of physics in 1903, until his retirement in 1925. Pierre's collaboration with Jacques, in which they shared their ideas constantly, prepared him to willingly do the same with **Marie Curie**. *See also* FAMILY OF P. CURIE.

CURIE, PIERRE (1859–1906). Pierre was born in Paris, the second son of Sophie-Claire Depouilly and **Eugène Curie**. His mother's father and brothers were commercial inventors, and both Eugène and his father were physicians. See the following entries concerned with different aspects of Pierre Curie's life and works: DEATH OF P. CURIE; EARLY RESEARCH, OF M. CURIE; EDUCATION OF P. CURIE; FAMILY OF P. CURIE; HEALTH PROBLEMS OF P. CURIE; POSITIONS, PROBLEMS WITH, FOR P. CURIE; RESEARCH OF P. CURIE.

CURIE, SOPHIE-CLAIRE DEPOUILLY (1859–1897). Sophie-Claire was the wife of Dr. **Eugène Curie**, the mother of **Pierre Curie**, and the mother-in-law of **Marie Curie**. Pierre Curie was very fond of his mother and devastated by her death two weeks after the birth of **Irène Curie**. *See also* FAMILY OF P. CURIE.

CURIE–LANGEVIN AFFAIR. Both **Marie** and **Pierre Curie** had long been associated with **Paul** and **Jeanne Langevin**, both professionally and socially. After Pierre's death, Marie continued the friendship with Paul Langevin. He confided in her the problems he and his wife Jeanne were having. Langevin and Curie had many common interests, and Marie, who was a private person, made the mistake of assuming that only a few close friends and colleagues would be interested in their relationship. Still, although their friendship was blossoming into love, they were careful to hide the affair. Marie did, however, speak of Paul Langevin's unhappiness to **Marguerite Borel**. In the spring of 1911, letters that Marie and Paul had written to each other were stolen by an unknown person who broke into Paul's study, and may have been hired by Jeanne Langevin. After the robbery, Madame Langevin and the newspapers were quiet. However, the relationship between Jeanne and Paul continued to deteriorate, and after an argument Paul took their two sons and left the house. Jeanne claimed that Paul had struck her in the face for cooking badly, and he countered that she had hurled insults at him in front of the boys. Because he had taken the children without Jeanne knowing where they were, he was subject to a lawsuit. He may have been paying blackmail to Jeanne in order keep the letters away from the newspapers. Marie loaned him money during this time. While Paul and Marie were in Brussels attending the

Solvay Conference, the formerly faint rumors blossomed into a scandal when *Le Journal* wrote a scathing article, which Curie and the article's author later repudiated.

Although for a time it seemed that the scandal would evaporate, Marie's name was constantly before the public. Curie had been nominated for her second **Nobel Prize**, this time in chemistry. Even though the Nobel committee accepted Curie's denial and that of the man who had written the article in *Le Journal*, the public remained involved. Scientist friends of both Curie and Langevin did what they could to support them, but both *l'Intransigeant* and *l'Action Française*, without the incriminating letters, castigated Curie and Langevin and insisted that Jeanne should have custody of the Langevin children. On 23 November 1911, *l'Oeuvre* acquired the actual letters and published incriminating excerpts from them. The hate campaign that followed almost resulted in Curie's not going to Sweden to accept the Nobel Prize. As the *New York Times* reported on 23 December 1911, the Langevin case had been dismissed. Jeanne Langevin was happy with the settlement of the divorce court, which granted her a separation from her husband and the custody of their four children. It also ensured that Paul was financially responsible for their maintenance. *See also* DUELS INVOLVING M. CURIE; *GIL BLAS*; *LE TEMPS*.

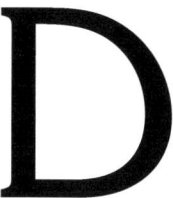

DAVY, SIR HUMPHRY (1778–1829). The chemist Sir Humphry Davy was born in Cornwall, England, and is best known for isolating a series of elements by electrolysis: aluminum, sodium, potassium, calcium, strontium, magnesium, barium, and boron. In doing so, he invented the new field of electrochemistry. He also discovered the elemental nature of chlorine and iodine. He was a mentor of the scientist Michael Faraday. He received many honors, including being named president of the **Royal Society**, secretary of the **Royal Institution**, a fellow of the Geological Society, and recipient of the Copley Medal, the Prix du Galvanism, the Rumford Medal, and the Royal Medal. The **Davy Medal** is named for him.

DAVY MEDAL. The Davy Medal, named for the chemist **Sir Humphry Davy**, is a bronze medal awarded annually by the **Royal Society of London** for an important recent discovery in any branch of chemistry. It was awarded to **Marie** and **Pierre Curie** in 1903 for their research on **radium**.

DEATH OF M. CURIE. *See* HEALTH PROBLEMS OF M. CURIE.

DEATH OF P. CURIE. Forty-six-year old **Pierre Curie** was killed in an accident on Thursday, 19 April 1906, when he was leaving a meeting of the Physics Society, where one of his passions, the teaching of physics, had been discussed. As he walked home in a cold drizzle about two in the afternoon through heavy traffic, he absentmindedly stepped in the path of a horse pulling a wagon. He slipped, fell, and was killed instantly when one of the wheels ran over his head, fracturing his skull. Before his death, Pierre had been suffering with what we now know as symptoms from his exposure to **radioactive** materials. His body was taken to a police station nearby, where they examined his papers and identified his body. When the crowd recognized who had been killed by the wagon, they turned upon its driver, who was protected by the police. **Marie Curie** was not home when Pierre's senior colleague and dean of the faculty of science, **Paul Appell**, and **Jean Perrin**, next-door neighbor and friend, went to the Curie house with the news of Pierre's accident. Pierre's father was alone at the house and received the news that Pierre had been killed. **Eugène Curie** was disconsolate and considered various reasons as to why it had happened, including daydreaming, slick streets, absentmindedness, or Pierre's illness. Marie returned home about six o'clock and immediately knew something was wrong. As Appell told the story, she did not cry, nor did she speak for several minutes. Her words when she finally spoke were "Pierre is dead? Absolutely dead?" Mme. Perrin cared for Irène at her house, but Eve remained at home cared for by others. When Marie spoke to Irène over the fence, she was unable to tell her daughter that her father was dead. Instead, she said that he had hurt his head badly in an accident and needed rest. *See also* PERRIN, HENRIETTE.

DEBIERNE, ANDRÉ-LOUIS (1874–1949). Debierne, a French chemist who discovered the element actinium, was a good friend of the Curies. Pierre was his doctoral advisor and requested that the young chemist work with them on **radioactivity**. He was not only a colleague of the Curies but also a good personal friend. After **Pierre Curie**'s death, he helped **Marie Curie** with her teaching and research and also served as a surrogate father to the children, especially Eve.

THE DELINEATOR. This women's magazine was edited by **Marie Mattingly Meloney**, known as Missy, who interviewed **Marie Curie**. Meloney managed to get an interview with Curie when she had turned away many others. *See also* UNITED STATES, M. CURIE'S 1921 TRIP TO THE; UNITED STATES, M. CURIE'S 1929 TRIP TO THE.

DEPOUILLY, SOPHIE-CLAIRE. *See* CURIE, SOPHIE-CLAIRE DEPOUILLY.

DLUSKA, BRONISLAWA (BRONIA) (1865–1939). Known as Bronia by her family, she was the second oldest of five children. She had three sisters—Zofia, Helena, and Maria (**Marie Curie**)—and one brother, Józef. Bronislawa graduated from secondary school with a gold medal. When it came to higher education in **Poland**, she faced the same problems that Maria and all scholarly inclined Polish women encountered. The universities did not admit women. She, like her sister Maria, joined the underground so-called **Floating University** to get what advanced education she could. She was able to save enough money through tutoring and accepting financial aid from Maria to go to Paris and study medicine at the **Sorbonne**. The two sisters agreed that Maria would help Bronislawa complete her medical studies and Bonislawa would reciprocate when Maria came to Paris. She graduated in 1890 as a gynecologist-obstetrician and married a fellow physician, **Kazimierz Dluski**. After graduation, she and Kazimierz ran a medical clinic, and their apartment became a meeting place for Polish exiles, immigrants, and expatriates. Bronislawa helped Maria when she came to Paris to study by hosting her for a short time. Maria, however, found Kazimierz annoying and the household so chaotic that she found it difficult to study, so she moved to a building closer to the university and lived in the garret. Bronislawa and Kazimierz had a daughter, Helena, and a son, Jakub. The Dluskis returned to Poland and set up a pulmonological sanitarium in the resort town of Zakopane. Kazimierz was not allowed to travel to the Russian part of the partitioned Poland. After the end of World War I, Poland regained its independence and Kazimierz joined the Polish delegation at the Versailles Peace Conference. The Dluskis could then return to **Warsaw**, and they set up a tuberculosis preventorium in a Warsaw suburb. Marie Curie had successfully opened a **radium institute** in Paris and worked toward setting up a second **radium** institute in Poland. Bronislawa supervised construction and recruitment while Marie raised funds in the United States and other places. The institute was officially opened in 1932 with Bronislawa as the first director.

DLUSKI, KAZIMIERZ (1855–1930). Dluski was a Polish physician who was the husband of **Bronislawa Dluska** and the brother-in-law of **Marie Curie**. A very social individual, he annoyed Marie when she was living in the Dluski house as a student because he often entertained people and the house was noisy and not conducive to Marie's studies. However, after Marie moved out and into her own garret, she became ill in her bitterly cold room and from a lack of nourishing food. Kazimierz brought her back to the Dluski house, where Bronia and he nursed her back to health.

DOCTORAL DISSERTATION OF M. CURIE. Marie Curie decided on the topic for her doctorate after pondering the significance of **Henri Becquerel**'s discovery of rays given off by **uranium** in March 1896. Her dissertation, she determined, would be an explanation of this phenomenon. However, many events delayed her actual writing of the dissertation, including ill health and the sickness and death of her beloved father. On 11 May 1903, the dean of the faculty, **Paul Appell**, approved the manuscript, and in June 1903, Maria successfully defended her dissertation. This document has been translated into many different languages.

DORABIALSKA, ALICJA (1897–1975). Alicja Dorabialska was a Polish chemist who received her Ph.D. degree from **Warsaw** University in 1922. She became skilled in research into **radioactivity** when she worked under **Marie Curie** in Paris at the **Radium Institute** in 1925. Later, Dorabialska wrote a book on the Curies and biographical articles on Marie Curie and other Eastern European scientists and coedited a Polish edition of Marie Curie's works. A prolific writer, she authored or edited over 150 works. She not only published her original research but also many review articles. Although Marie Curie had a difficult time acknowledging that her **radium** could cause physical harm, she confided in Dorabialska that, although she did not fully understand the effect of radium on health, she feared it might be the cause of her cataracts and the reason for her uncertain gait. *See also* HEALTH PROBLEMS OF M. CURIE.

DREYFUS, ALFRED (1859–1935). Alfred Dreyfus symbolized the evil part of the climate of **Third Republic** Paris when the Curies were working there. French anti-Semitism was rife when in 1894 Dreyfus, a French Jewish artillery officer, was falsely convicted of spying for Germany. Dreyfus's trial and conviction, now known as the Dreyfus affair, resulted in a sentence of life imprisonment on Devil's Island off the coast of South America. Even after he was eventually exonerated, the repercussions of the affair were felt over France and the rest of Europe.

DUELS INVOLVING M. CURIE. Although five duels were reported between those who took **Marie Curie**'s side and those who supported Mme. Langevin, three were the most significant. The first duel was between her supporter Henri Chervet (1881–1915) of *Gil Blas* and detractor Léon Daudet (1867–1942) of *l'Action Française*. They fought with swords, and Daudet, although the more experienced dueler, suffered a deep wound to the elbow. The second duel, between *Gil Blas* writer **Pierre Mortier** (1882–1946), a Curie supporter, and **Gustave Téry** of *l'Oeuvre*, resulted in Mortier being wounded in the arm. Téry had insulted **Paul Langevin**, resulting in Langevin challenging Téry to a duel. This duel, with pistols, ended with neither combatant firing. When it came down to the actual firing, Téry demurred, realizing, as he mentioned later in *Gil Blas*, that by killing Langevin he would be depriving France of one of its most famous scientific minds. *See also* CURIE–LANGEVIN AFFAIR.

E

EARLY RESEARCH OF M. CURIE. During the time that Maria Sklodowska was completing her mathematics degree, she was hired by an organization with a goal of promoting French science, the Society for the Encouragement of National Industry. Her assignment was to study how the magnetic properties of various steels varied with their chemical composition. She was hampered by the lack of research laboratory space. A Polish couple, physicist Józef Kowalski and his wife, whom Maria had met when she was a **governess**, were in Paris for their honeymoon and heard of her need. They suggested that she meet with their friend **Pierre Curie**, who was working on magnetism at the nearby **École de Physique et Chimie Industrielles** and possibly had some laboratory space that she could use. Perhaps the Kowalskis had a romance in mind, but certainly neither Pierre nor Maria had any idea of such. The meeting resulted in Pierre providing Maria with research space. She was supplied with free samples and had the advice of both Pierre and the chemist **Henri Le Châtelier**. The paper that resulted was completed in the fall of 1897 and, though it lacked originality, provided her with experience for her next, more creative project.

ÉCOLE DE PHYSIQUE ET CHIMIE INDUSTRIELLES DE LA VILLE DE PARIS (ESPCI) (SCHOOL OF INDUSTRIAL PHYSICS AND CHEMISTRY) (FOUNDED 1882). After this school's establishment, it became a meeting place for the country's best scientists. **Pierre Curie** and his brother **Jacques Curie** collaborated on a study of the electrical properties of crystals that led to the discovery of **piezoelectricity** at the **Sorbonne**. Their major collaboration ended in 1883, when Jacques accepted a position in mineralogy at the University of Montpellier. Pierre then left his position at the Sorbonne and accepted a position at the then new School of Industrial Physics and Chemistry. This school was located in old crumbling buildings, and the laboratories were far inferior to those at the Sorbonne. However, Pierre honed his skills as a teacher and was happy with the freedom that he experienced there. He left ESPCI and applied for a vacant chair in physics at the Sorbonne but was turned down. He eventually was hired by that institution as a member of the faculty, teaching physics to medical students. Even though this made him a member of the faculty at the Sorbonne, since the course served medical students it was peripheral to the prestigious Faculté des Sciences, where Pierre wanted to be. During the early years of their marriage, **Marie Curie** received permission to work at the ESPCI with Pierre, although she had to finance her own research. Her project resulted in the completion of her first paper in the fall of 1897, which examined the magnetic properties of difference tempered steels and how these varied with their chemical composition. She completed her first paper on this subject in 1897. Although she realized that it was not an especially novel paper, it allowed her to work in a laboratory, and, as she said, it was better than giving lessons to students. *See also* EARLY RESEARCH OF M. CURIE; RESEARCH OF P. CURIE.

ÉCOLE NORMALE SUPÉRIEURE DE JEUNES FILLES. This institute of higher education for girls only in Sévres was founded in 1881 and existed until 1985, when it merged with the École Normale Supérieure to form a coeducational school. **Marie Curie** was the first woman to teach at this school, known as the best preparatory school for women teachers in France. Her salary was necessary at this time because in 1900 the family had moved into a larger, more expensive house in the outskirts of Paris in order to accommodate **Eugène Curie**, who was living with them and caring for the children. **Pierre Curie** had also taken a second job in order to make ends meet. When Marie first began to teach, her inexperience was evident and her students complained about her Polish accent and awkward syntax. However, after the first year, she improved greatly by using experiments and discussing the implications of the experiments with her students instead of lecturing. After Pierre's death and Marie became the first woman to teach at the **Sorbonne**, she had to leave her job at Sévres precipitously, and she arranged for her students to attend her lectures at the Sorbonne.

ÉCOLE POLYTECHNIQUE (FOUNDED 1794). After **Pierre Curie** had been passed over for the chair of physical chemistry at the **Sorbonne** in 1898, he settled on a post as assistant professor at the École Polytechnique in order to try to supplement their income. The Curies were disappointed and almost left Paris for Geneva. In Geneva, he was promised a chair with a high salary, a laboratory that he could design himself, and an official position for Marie. After first accepting the position in Geneva, the Curies changed their mind, fearing it would cause an interruption in their investigations. Fortunately, another position opened at the Sorbonne, teaching physics to medical students. Although it was not an ideal position, it helped with the finances.

EDUCATION OF M. CURIE. *See* SORBONNE, M. CURIE AS A STUDENT AT THE; STUDENT, M. CURIE AS A PREUNIVERSITY.

EDUCATION OF P. CURIE. Pierre was homeschooled by his father and older brother, **Paul-Jacques Curie** (usually referred to as "Jacques"), until he was 14, when he was tutored by a professor of mathematics, Albert Bazille. Until this tutor came on the scene, Pierre was not considered a particularly fast learner; in fact, he himself thought he had a "slow mind." But Bazille inspired him to think creatively. When he was 16, he passed the examination that allowed him to enter the **Sorbonne**, and at 18 he received the *licence ès sciences*. He then became an assistant in the Sorbonne physics laboratory, where students carried on experiments. He soon became skilled enough to publish original work, sometimes in collaboration with a teacher and later with his brother, Jacques, who was a laboratory assistant in the laboratory of mineralogy at the Sorbonne. Pierre was undoubtedly not a traditional learner, nor was he interested in academic advances.

EINSTEIN, ALBERT (1879–1955). German-born theoretical physicist Einstein was undoubtedly one of the most important scientists of all time. He also was important in the life of **Marie Curie**. Einstein, best known for his two theories of relativity, had an approach to science completely different from that of Marie Curie. Whereas Curie was a persistent laboratory scientist, Einstein never engaged in laboratory work. He worked with his mind not his hands, but understood the importance of Curie's approach. Recognizing that classical Newtonian mechanics failed to provide a credible explanation for James Clerk Maxwell's equations describing the electromagnetic field, Einstein developed his special theory of relativity (1905) while working at the Swiss patent office in Bern, Switzerland, before he met Marie Curie. This theory demonstrated mass–energy equivalence ($E = mc^2$). The year 1905 was considered his *annus mirabilis* because of the four important papers he published in that year including special relativity.

In 1917, Einstein published his general theory of relativity, in which he refuted the long-held Newtonian view on gravity. He stated that an apparent gravitational attraction between mass is the result of the warping of space–time by these masses. Einstein's views on these subjects and his many other scientific ideas were despised by the anti-Semitic Nazi regime. Realizing that living in Germany was no longer an option, Einstein came to the United States and took a position at the Institute for Advanced Study at Princeton University and became a U.S. citizen in 1944. Einstein was a well-known supporter of the Zionist movement, had complex philosophical and religious views, and although he was basically a pacifist, when he realized the possibility that Germany might win a race to build the atomic bomb, he signed a letter to President Roosevelt supporting the possibility of building an atomic bomb, which he later considered a terrible mistake.

Einstein's experience with Marie Curie may have begun in 1911, when they both attended the **Solvay Conference** in Brussels. Although Curie did not present a paper, she engaged in informal socializing between sessions and impressed Einstein with her ability. However, Curie's career was put on hold after the conference because **Paul Langevin**'s wife had given the newspapers copies of love letters between Curie and Langevin. Cries of home wrecker from the populace threatened Curie's reputation. She became ill both physically and psychologically thereafter, and her scientific work was on hold. Einstein was very supportive of Curie and was enraged by the public's response. He wrote a supportive letter to Curie, which was just the beginning of a long correspondence between the two. Einstein and his first wife, **Mileva Marić Einstein**, visited Curie in 1913. They made plans to vacation together hiking in the Swiss Alpine passes. The group included Einstein's son **Hans Albert Einstein**, Curie's two daughters, and the Curie daughters' governess but not Mileva. On this trip, Einstein, who had divorced his first wife and was courting his cousin Elsa, noted that Curie expressed her feelings by grumbling and was, as noted by others, dour and dreary. In spite of personality differences, the two remained friends throughout their lives. After Curie's death, Einstein praised her while noting that she was an "unusually independent character," implying that it sometimes made it difficult for her to change her mind if she was certain that she was correct ("Mme Curie Is Dead; Martyr to Science," *New York Times*, 5 July 1934).

EINSTEIN, HANS ALBERT (1904–1973).

Hans Albert was the second child and first son of **Albert** and **Mileva Einstein**. Born in Switzerland, he came to the United States and became professor of hydraulic engineering at the University of California, Berkeley, where he was known for research on sediment transport. He and his wife, Frieda, had only one biological child who survived and produced children, Bernard Caesar. They had two other sons who died in infancy or childhood and one adopted daughter. As a boy, Hans Albert joined Einstein and **Marie Curie** and her daughters on a hiking vacation in the Swiss Alps in 1913.

EINSTEIN, MILEVA MARIĆ (1875–1948).

The first wife of **Albert Einstein**, Mileva Marić was a Serbian mathematician who was a fellow student with Albert at Zürich's Polytechnic and

Marie Curie and Albert Einstein

had a major influence on his scientific accomplishments. The extent of their collaboration is unclear. Marić was the mother of Albert's three children. She and Albert separated in 1914 and divorced in 1919.

ELECTROMETER. The instrument invented by the Curie brothers to measure small electric currents. *See* CURIE, PAUL-JACQUES; CURIE, PIERRE; RESEARCH OF P. CURIE.

ELECTRON (DISCOVERED 1897). The negatively charged subatomic particle important in **Wilhelm Röntgen**'s discovery of X-rays was first discovered by **J. J. Thomson** when he was studying the properties of **cathode** rays, which emit electrons. **Marie Curie** was familiar with the scientific literature and was aware of Röntgen's discovery of X-rays. **Henri Becquerel** was interested in the phenomenon of **phosphorescence** and proceeded to search for different rays emitted by different substances. In his search, Becquerel actually discovered **radioactivity** but did not name it or explain its source. He continued to believe in a hypothesis of phosphorescence. Curie was intrigued by Becquerel's new ray as she searched for a subject for her **doctoral dissertation**.

FAMILY OF P. CURIE. Pierre Curie's father was French Huguenot physician **Eugène Curie** (1827–1910), and his mother was **Sophie-Claire Depouilly Curie** (1832–1897). Pierre had an older sibling, **Paul-Jacques Curie**, known as Jacques. Although the two siblings were very close and collaborated on scientific work, their personalities were very different, with Jacques being much more assertive and often arguing with their father. Pierre married Maria Sklodowska in 1895, and the couple had two children, Irène and Eve.

FLOATING UNIVERSITY. While **Poland** was controlled by Russia during **Marie Curie**'s youth and women were denied a university education, an underground institution known as the Floating University flourished. This "university" had a regular curriculum and met for two hours a week to discuss such revolutionary ideas as **Marxism** and **positivism**. Maria and her sister Bronia both attended but were dissatisfied with the ad hoc attempts at education and wanted to go abroad. *See also* COMTE, AUGUSTE.

FRACTIONAL CRYSTALLIZATION. When **Marie Curie** was attempting to separate different substances from **pitchblende** in search of a new highly radioactive material, she tried several different methods. One of these techniques was called fractional crystallization. It is based on the observations that different substances in the same solution form crystals at different temperatures. Substances with lower atomic weights crystallize first. Curie first boiled the pitchblende solution and then gradually cooled it and tested the crystals for **radioactivity**. She discarded the first-formed crystals that were not radioactive or only slightly so. She continued by repeating this technique over and over again on the solution. She retained the more radioactive fraction and discarded the less active crystals. With each fractional crystallization, the crystals became increasingly more active. This process was exceedingly tedious, and it still did not yield the desired result. The element bismuth stubbornly refused to be separated from the supposed new element in the pitchblende.

FRANCO-PRUSSIAN WAR (1870–1871). The French parliament voted to declare war on Prussia on 16 July 1870, and the war began three days later when France invaded Prussian territory. The Prussian army was far superior to the French army and decisively defeated the army of the Second Empire and captured the Emperor Napoleon III. France declared the **Third Republic** on 4 September 1870, and continued the war for another five months, when the French forces were defeated in northern France. Paris fell on 18 January 1871, and a revolutionary uprising called the **Commune of Paris** seized power for two months until it was finally suppressed by the French army in May 1871. These events caused the French universities to no longer be superior. The sciences especially suffered in comparison to the German universities, where current research was discussed and students were trained in laboratory work. After the war, the French saw the need to reform their educational institutions, especially their science faculties. By the time **Marie Curie** attended the **Sorbonne**, one of the oldest universities in the world, it had begun to make advances in the science curriculum although the German universities still outperformed the French. Curie approved of the Sorbonne's new emphasis on republican anticlerical teachings. The humanities were de-emphasized and the sciences stressed, even to the point of a massive building project with new science classrooms and laboratories as well as hiring superior science teachers. This new emphasis was important to Marie Curie's success.

FRENCH ACADEMY OF SCIENCE. *See* ACADÉMIE DES SCIENCES.

FULLER, LOIE (1862–1928). Loie Fuller was an American actress and dancer who was important in the development of modern dance and theatrical lighting. After **Marie Curie**'s daughter Eve was born and Marie had received her first **Nobel Prize**, her life was very busy with home, children, husband, teaching, and research. Nevertheless, during this time in her life, she found more time than previously to socialize with friends. Loie Fuller had made a great impression on Parisians with her dancing with veils illuminated by colored lights. Marie met her when she presented a special show for the Curies at their house. They remained friends for many years, and she was supportive when Curie was under siege during the **Paul Langevin** affair. When Fuller developed breast cancer, she asked Curie's advice about the use of **radium** as a treatment. Curie referred her to the director of the medical division of her **Radium Institute**. *See also* CURIE–LANGEVIN AFFAIR.

GEGNER PRIZE. This prize, given by the **Académie des Sciences**, was awarded to **Marie Curie** in 1903 for scientific promise.

GEIGER, HANS (1882–1945). Ernest Rutherford at McGill University devised an experiment that led to the idea that the atom could be cut into another part in addition to the **electron**. He suggested an experiment to his colleagues Hans Geiger and **Ernest Marsden**. They shot positively charged **alpha particles** at a thin sheet of gold, assuming that the particles would go straight through the foil with very little deflection. The accepted theory at that time was the "plum pudding model" of the atom. The negative electrons (plums) would be spread evenly throughout the positive matrix (the pudding). They were surprised to find that although most of the positive particles went straight through the foil (98 percent), a small percentage were deflected at large angles (about 2 percent), and .01 percent bounced off the gold foil. Since alpha particles have about 8,000 times the mass of an electron, it was obvious that strong forces were necessary to deflect the particles. They counted the alpha particles, and Rutherford interpreted the results to mean that the mass of an atom was concentrated into a compact positive nucleus with electrons occupying most of the atom's space. Geiger is probably best known as the inventor of the Geiger counter.

GIL BLAS **(FOUNDED 19 NOVEMBER 1879).** A Parisian literary periodical named for Alain-René Lesage's novel *Gil Blas*, this periodical was founded by the sculptor Augustin-Alexandre Dumont. During the **Curie–Langevin affair**, a supporter of **Marie Curie**, Henri Chervet of *Gil Blas*, fought a duel with her detractor Leon Daudet of *L'Action Française*. This was the first of five duels to be provoked by the affair. *See also* DUELS INVOLVING M. CURIE.

GIRTON COLLEGE, CAMBRIDGE (ESTABLISHED 1869). This first residential college for **women** at Cambridge University was established by Emily Davies, **Barbara Bodichon**, and Lady Stanley of Alderly. Located about 2.5 miles northwest of the city of Cambridge, where the university with the men's colleges were located, it was less of a threat to the venerable male colleges than if it were located near them. Although the university did not allow the women of Girton to obtain degrees in the 19th century, the female students were allowed to sit for the examinations. Girton was granted full college status in 1948. Marie Curie's friend **Hertha Ayrton** was befriended by Barbara Bodichon, who collected enough money to allow her to enter Girton.

GLEDITSCH, ELLEN (1879–1968). Chemist Ellen Gleditsch graduated at the top of her high school class in Mandal, Norway, but since **women** were not allowed to take the entrance examination for academic degrees she did not immediately work toward a university degree. Instead, she became a pharmacy assistant and received a nonacademic degree in chemistry and pharmacology. With the aid of a mentor, she eventually was allowed to take the university entrance examination and continued her education at the University of Oslo. She chose to leave Norway and decided to study **radioactivity** in Paris at the **Sorbonne**. Recommended by a professor, she gained entrance to **Marie Curie**'s laboratory, where she worked from 1907 to 1912. One of her tasks was performing **fractional crystallizations** to purify **radium**. After testing, she refuted British chemist **William Ramsay**'s claim that radiations from radioactive substances could cause transmutation. To do this, she determined the radium–**uranium** ratio in minerals of varying age and origin. She found significant differences in the ratios, but they were not large enough to exclude radium's descent from uranium. To explain the discrepancy, she postulated the presence of a long-lived intermediate product. This hypothesis was later confirmed. She determined radium's period and decay constant. After five years in Curie's laboratory, she received a *licence en sciences* degree from the Sorbonne and left Paris for a teaching position at the University of Oslo. She maintained contact with Curie, returning at intervals to work in her Paris laboratory. She left Oslo to work in the laboratory of **Bertram Boltwood** at Yale University. While there, she created a standard measurement for the half-life of radium. She received an honorary doctorate at Smith College for her work. She also was the first woman to receive an honorary doctorate from the Sorbonne. She published numerous articles and several books on radioactivity. She also authored a textbook of inorganic chemistry and a biography of the French chemist Antoine Laurent Lavoisier (1743–1794).

GOVERNESS, M. CURIE AS A. Since **women** were not allowed to attend Polish universities when Maria Sklodowska was ready for higher education, and because her father was financially unable to send his daughter to a foreign university with more lenient attitudes toward women scholars, she decided to take a governess position in order to save money to study abroad. Her first job in **Warsaw** was a disaster, and she soon resigned. Nevertheless, she realized that the only way that she could get an education was by working as a governess. She and her sister Bronia agreed on a plan where each would get the education that she wanted. Maria would find another governess position, this time outside of Warsaw, and save most of her salary. Since Bronia was the older, she would be the first to benefit. She would attend medical school in Paris and when she competed her degree would help Maria. Maria's second attempt as a governess was more successful. She described her employers, the Zorawskis, as "excellent people." Although Bronka, the older daughter, and Maria were about the same age, Maria was better educated. The two girls got along very well, but the parents made it evident that Maria, as a governess, was inferior in status. When a romance developed between Maria and **Kazimierz Zorawski**, the older son of the family, the parents did all in their power to break up the relationship. It was unimportant to the Zorawskis that Maria was intelligent, came from a good family, and was obviously a refined person; she was still an employee, a lowly governess without money. They successfully broke up the romance. Nevertheless, Maria remained as governess for 15 more months but was very unhappy. During the meantime, sister Bronia had become engaged to a fellow medical student. After they were married, she invited Maria to come to Paris and stay with them. When she left the Zorawskis, she returned to Warsaw, stayed with her father, and took another position as governess with the Fuchs family. Although this job did not have the problems that she had with the previous ones, she was still unhappy. However, her father became better off financially and she, for the first time, had access to a small laboratory run by her cousin, **Józef Boguski**, codifying her interest in experimental physics and chemistry. *See also* DLUSKA, BRONISLAWA; WOMEN; ZORAWSKI FAMILY.

GYMNASIUM. The gymnasium in the German education system is the most advanced of German secondary schools. Gymnasia can be public and state funded or parochial or private. The gymnasium that Maria Sklodowska

(**Marie Curie**) attended, **Gymnasium Number Three**, was run by the government. *See also* STUDENT, M. CURIE AS A PREUNIVERSITY.

GYMNASIUM NUMBER THREE. Marie Curie attended this Russian-dominated advanced secondary school. This public **gymnasium**, although based on the German prototype, suffered from a lack of good teachers, many of whom were chosen because of political correctness and others who were merely incompetent. Nevertheless, after Maria's mother died, she was taken out of the nurturing school of **Jadwiga Sikorska** and placed in what to Maria seemed a hostile environment. Maria's father knew the problems with the Polish gymnasia because he taught in a male-only school where Maria's brother, Józef, was a student. However, he insisted that Maria attend. In Józef's gymnasium, although the Polish language was forbidden outside of the actual classes, both students and teachers surreptitiously spoke Polish. However, in Maria's Gymnasium Number Three, the enforcement of the Polish language was strictly adhered to. However, when Maria attended the school, it had some excellent teachers, including the physics teacher. Although she complained about the school later in her life, closer to the time she attended she confessed to actually liking school. She graduated from Gymnasium Number Three in 1883, finished first in her class, and was awarded the gold medal. *See also* STUDENT, M. CURIE AS A PREUNIVERSITY.

HARDING, WARREN (1865–1923). Harding was the 29th president of the United States (4 March 1921–2 August 1923). He was a Republican and very popular during his campaign and into his presidency. However, after his death, subsequent disclosures of scandals that permeated his administration such as the Teapot Dome scandal eroded his popular appeal, and he is often ranked as the worst of the American presidents. His political career before he was elected president was as a member of the Ohio Senate (1900–1904), lieutenant governor of Ohio (1904–1906), and United States senator from Ohio (1915–1921). He died of a heart attack while on a speaking tour and was succeeded by his vice president, **Calvin Coolidge**. During **Marie Curie**'s first trip to the United States to receive a gram of **radium** from funds raised by American women inspired by **Marie Meloney**, she was invited to the White House, where President Harding presented her with a gram of **radium** (although it actually was a facsimile). Harding referred to her as a noble creature, devoted wife, and loving mother. He also referred to her as foremost among scientists and praised her for being a leader among **women**. *See also* UNITED STATES, M. CURIE'S 1921 TRIP TO THE.

HARVARD COLLEGE OBSERVATORY. During the 19th and early 20th centuries, astronomy was one of the few fields where **women** could get paying positions outside of the home. Nevertheless, the positions open to women were ones that men had shunned. The adoption of technologies such as cameras and spectroscopes opened up a new labor market for women. **Edward Pickering**, the director of the Harvard College Observatory, hired women computers to classify as cheaply as possible the thousands of photographic plates generated by his new equipment. Low-paid positions such as these were important in convincing people that women could make significant contributions to science.

HEALTH PROBLEMS OF M. CURIE. Throughout her lifetime, **Marie Curie** suffered problems with her health, some of which were probably stress related. She became so obsessed with her work or problems with her private life that she did not take care of herself. As early as 1897, both **Pierre Curie** and Marie had begun to have serious health problems. They, as well as their friends, blamed their problems on overwork and not eating or sleeping properly. Marie was diagnosed with a possible tubercular lesion of the lung, although it did not progress beyond the initial symptoms. The timing of this illness corresponds with the heavy doses of radiation that she was exposed to. Marie kept an accurate account of her symptoms in her notebook. Although it was understood that **radium** exposure could cause local burns, the systemic effects of radium were unknown. In 1903, the year that she, Pierre, and **Henri Becquerel**, won the **Nobel Prize** in Physics, she had a miscarriage after one of their long **bicycle** rides. Although she had been exposed to very high doses of radiation during this pregnancy, she did not attribute to it the loss of the baby. After the miscarriage, Marie soon became pregnant again, and this time the result was a healthy baby girl, Eve Denise.

After she gave her second Nobel lecture, Marie became seriously ill with a severe kidney ailment. Her close friends thought that it was precipitated by the fallout from the **Paul Langevin** affair. Although the acute infection went away and she went back to work in her laboratory, physicians still recommended surgery, which she had in March 1912. The surgery to remove the lesions was a success, but her health was compromised for several months. During this time, she fell into a deep depression, so much so that her friends feared she might take her own life. Around 1921, she developed cataracts. Although she was suspicious that radium was the cause, she attempted to keep her illness secret. She admitted to her sister Bronia that her eyes were weaker and she had a continuous humming in her ears. In 1932, Marie had an accident in the laboratory resulting in a broken right wrist.

This accident seemed to be the beginning of a rapid decline in her health. It seemed to have brought to the surface the health problems that had been dormant, such as inflamed radiation burns and extreme headaches. She was confined in her bed for long periods of time. In December 1933, she had a large stone in her gall bladder but resisted surgery. Even though she did not admit, even to herself, that her health was abysmal, even going skating and skiing in the winter of 1934 with Frédéric and the children, at some level she probably knew that her death was imminent. She explained to Irène where she could find the documents that would serve as her will for the gram of radium along with some other papers. She destroyed all of the personal documents, especially those she found painful. In 1934, her health continued to deteriorate and she became unable to drag herself into the laboratory. The doctors' diagnosed her as having a recurrence of tuberculosis and suggested a stay at a sanatorium. By the time they reached the sanatorium (where she insisted on registering with a pseudonym, Madame Pierre, to protect her privacy), her temperature had spiked and both her red and white blood cell counts fell. On 3 July, Marie was convinced that she was improving, noting that her temperature had fallen. Eve, however, who had accompanied her mother to the sanatorium, realized as did the doctors that a decrease in temperature often precedes death. Before her death, she began to hallucinate, and on 4 July 1934, Marie Curie died at the **Sancellemoz Sanatorium**. The attending physician diagnosed her disease as aplastic pernicious anemia, probably caused by a long accumulation of radiation. *See also* CURIE–LANGEVIN AFFAIR; JOLIOT-CURIE, FRÉDÉRIC.

HEALTH PROBLEMS OF P. CURIE. By 1897, both **Pierre** and **Marie Curie** were having health problems, but they blamed them, as did their friends, on overwork, lack of sleep, and improper nutrition. Although they did not acknowledge it, the timing seemed to indicate that the new rays they were investigating had something to do with their ill health. In 1901, Pierre and **Henri Becquerel** published a paper in which they described burns on their skin caused by contact with **radioactive** material. Becquerel was burned while carrying a tube of **radium** in the pocket of his waistcoat. Two Germans reported the first incident of burns on their skin caused by radioactive materials. Pierre duplicated their experience by placing thickly wrapped barium on his arm for 10 hours. His skin became red, and the redness increased for several days. The area began to heal around the edges on the 42nd day, and by the 52nd day it was all healed except for a small gray spot. Rather than worry about this phenomenon, and the burns both he and Marie had on their hands and fingers, they dismissed the burns as minor problems. In fact, they saw a possible positive benefit from it. Perhaps radium could be used to destroy diseased cancer cells. In 1902 and 1903, Pierre's general health became increasingly fragile. When he presented a lecture at the **Royal Institution in London** and accidentally spilled a tiny quantity of radium on the table, 50 years later the level of radioactivity was such that the building required decontamination. By 1906, Pierre's health had gotten decidedly worse. Although radiation wasn't the immediate cause of Pierre's death on 19 April 1906, it may have influenced his absentmindedness as he walked home from his meeting. *See also* DEATH OF P. CURIE; HEALTH PROBLEMS OF M. CURIE.

HOOVER, HERBERT (1874–1964). Herbert Hoover was the 31st president of the United States (4 March 1929–4 March 1933). He was a Republican who succeeded **Calvin Coolidge** and was president at the beginning of the Great Depression. During **Marie Curie**'s second visit to the United States in the fall of 1929, President Hoover presented her with a check raised by a group of American women for **radium** for her Polish institute. Curie was Hoover's guest in the White House during this visit. *See also* UNITED STATES, M. CURIE'S 1929 TRIP TO THE.

HUGGINS, MARGARET LINDSAY (MURRAY) (1848–1915). Astronomers Margaret and **William Huggins**, like the team of **Marie** and **Pierre Curie**, illustrate the importance of a husband–wife collaboration. Like Marie Curie, Margaret became interested in science before she met her husband. Her amateur astronomer grandfather spent evenings teaching Margaret how to identify the constellations. A common interest in spectroscopy brought Margaret and William Huggins together. When she married William, she stepped into a ready-made set of astronomical problems. Although she is usually considered his assistant, skilled at photographic manipulations and visual observations, William himself considered her work significant on its own and added her name as a coauthor in their later publications. The Hugginses' careers represent an earlier manifestation of the importance of marital collaboration to the history of science.

HUGGINS, WILLIAM (1824–1910). When William Huggins married his wife Margaret (26 years younger), he had already recognized the potential of several new observational techniques. After studying **Gustav Kirchoff**'s ideas on spectroscopy that the lines in the solar spectrum showed the chemical composition of its atmosphere, Huggins extended his methods to the stars. In doing so, he rescued the nebular hypothesis for the origin of the stars from obscurity. When he married Margaret, he had a new partner in his observational work in spectroscopy. **Margaret Huggins** was often characterized as William's assistant, skilled at photographic manipulations and visual observation, but later on he added her name to their publications. However, he was always listed as the senior author and made it clear that this was the case. In the case of the collaboration of the Curies, Pierre credited Marie with her work, and of course she also published under her own name. Nevertheless, the partnership of the Hugginses' was an earlier illustration of the importance of marital collaboration.

I

INDEPENDENT (25 JUNE 1903). An article in an issue of an American popular journal described **Marie Curie**'s investigations on **radioactive** substances. It praised her work and helped establish her reputation in the United States. It also expressed wonder that a **woman** had been associated with the discovery of **radium**.

INSTITUT DU RADIUM. See RADIUM INSTITUTE.

INTELLECTUAL CLIMATE OF PARIS DURING M. CURIE'S STUDENT DAYS. When Marie Sklodowska arrived in Paris as a student, she was introduced to a vibrant intellectual society. During the turn of the century, this city was the artistic capital of the world. However, the dour **Marie Curie** and her idealistic husband remained mostly aloof from the popular culture that characterized the Montmartre section of Paris, where Toulouse-Lautrec (1864–1901) flourished at the center of cabaret entertainment and Bohemian life. Even though as a devoted student Marie had little time for cultural experiences, she could not fail to be influenced by the plethora of music and art that surrounded her. For instance, the musician Claude Debussy (1862–1918) represented impressionism in music, including the *Prelude to the Afternoon of a Faun* (*Prelude a l'apres-midi d'un faune*). The artist Edgar Degas (1834–1917), best known for his depictions of dancers, took advantage of the intellectual climate and painted in many different styles. The impressionist artists Claude Monet and Pierre-Auguste Renoir (1841–1914) are examples of the many cultural icons who made Paris of the late 19th and early 20th centuries an exciting an innovative place to live and work.

INTERNATIONAL BUREAU DES POIDS ET MESURES (BIPM). See INTERNATIONAL BUREAU OF WEIGHTS AND MEASURES NEAR PARIS.

INTERNATIONAL BUREAU OF WEIGHTS AND MEASURES NEAR PARIS. The bureau was created on 20 May 1875, and had a mandate to provide the basis for a single, coherent system of measurements throughout the world. Member states act together on matters related to measurement science and measurement standards.

INTERNATIONAL STANDARD FOR RADIUM. See RADIUM, INTERNATIONAL STANDARD FOR.

L'INTRANSIGEANT. This French newspaper founded in July 1880 by Henri Rochefort originally represented the left-wing opposition but moved to the right. By the time **Marie Curie** was a candidate for the French Academy of Science, it was a strident right-wing publication that accused Marie Curie of being little more than a hack. Their opinion was that it was ridiculous to consider a **woman** for membership in this male institution. During the **Curie–Langevin affair**, it took Madame Langevin's side and berated Marie Curie at every opportunity. It ceased publication after the French surrender in 1940. *See also* ACADÉMIE DES SCIENCES.

J

JOACHIMSTHAL REGION. A region on the German–Czech border where **pitchblende**, which contained **uranium**, was mined. The region was first famous for its silver mines. This region is where the Curies extracted uranium from large quantities of this substance. *See also* KLAPROTH, MARTIN HEINRICH.

JOLIOT, FRÉDÉRIC. *See* JOLIOT-CURIE, FRÉDÉRIC.

JOLIOT-CURIE, FRÉDÉRIC (1900–1958). Physicist Frédéric Joliot was born in Paris and as a young man studied at the École Supérieure de Physique et de Chimie Industrielles de la Ville de Paris, where he graduated first in his class and became an early admirer of the Curies. **Paul Langevin** was his mentor and suggested that he work in **Marie Curie**'s laboratory. Marie immediately accepted him, and **Irène Joliot-Curie**, who was several years older and was experienced in the laboratory procedures, tutored him in the proper techniques. Marie Curie was surprised when she was informed of the impending marriage of Frédéric Joliot and her daughter Irène. When first hearing of the proposed match, Marie feared that the difference in the personalities of Frédéric and Irène would result in Irène being hurt. Frédéric was outgoing, whereas Irène was socially inept and somewhat forbidding. However, Marie was soon won over when she recalled her relationship with Pierre.

JOLIOT-CURIE, HÉLÈNE. Hélène was the daughter of **Irène Joliot-Curie** and Frédéric Joliot-Curie. *See* LANGEVIN-JOLIOT, HÉLÈNE.

JOLIOT-CURIE, IRÈNE. The older daughter of **Marie** and **Pierre Curie** and the wife of **Frédéric Joliot-Curie**, she and her husband were jointly awarded the **Nobel Prize** in Chemistry in 1935 for their discovery of artificial **radioactivity**. The couple had two children, Hélène Langevin-Joliot and Pierre Joliot-Curie. See the following entries concerned with different

Irène Joliot-Curie and Marie Curie

aspects of her life and works: BIRTH AND INFANCY OF I. JOLIOT-CURIE; CHILDHOOD AND EDUCATION OF I. JOLIOT-CURIE; MARRIAGE OF IRÈNE CURIE TO FRÉDÉRIC JOLIOT; NOBEL PRIZE FOR CHEMISTRY, 1935, I. JOLIOT-CURIE AND F. JOLIOT-CURIE; WORLD WAR II, EXPERIENCE OF I. JOLIOT-CURIE IN.

JOLIOT-CURIE, JEAN FRÉDÉRIC. *See* JOLIOT-CURIE, FRÉDÉRIC.

JOLIOT-CURIE, PIERRE (B. 1932). Pierre Joliot-Curie was the second child of **Irène** and **Frédéric Joliot-Curie**. Named after his famous grandfather, **Pierre Curie**, his great grandfather **Jacques Curie** was pleased that the name would remain in the family.

(LE) JOURNAL. Le Journal, a famous French newspaper published from 1892 to 1944, launched a front-page story headlined "A Story of Love: Madame Curie and Professor Langevin." **Jeanne Langevin** and her mother supplied material to the paper, and Jeanne came off as the wronged wife. **Marie Curie** was portrayed as a harridan who was stealing another woman's husband and was spoiling the lives of their children. This story came out while Marie and **Paul Langevin** were in Brussels attending the first **Solvay Conference** on radiation (1911). When Marie returned, she wrote a scathing denial and threatened to demand monetary damages to be used to promote science; the *Journal* reporter retracted the story, but the accusations did not die. *See also* CURIE–LANGEVIN AFFAIR.

K

KELVIN, LORD (WILLIAM THOMSON) (1824–1907). William Thomson, Lord Kelvin, was born in Belfast, but after his father was appointed professor of mathematics at Glasgow the family moved to that city, where he attended school. Later, he studied at Peterhouse College, Cambridge, and graduated as Second Wrangler and later became a fellow at that college. When the chair of natural philosophy (physics) opened at the University of Glasgow, Thomson was elected to fill it. He remained at Glasgow for the rest of his life. He was known for his work in thermodynamics, including playing an important role in the development of the second law of thermodynamics, the absolute temperature scale measured in kelvins, the dynamical theory of heat, the mathematical analysis of electricity and magnetism, the determination of the age of the earth, and work in hydrodynamics. He also became involved in technology and engineering, and his inventions were important in developing submarine cables.

Kelvin became interested in the work of the Curie brothers, Pierre and Jacques, on **piezoelectricity**. In a series of papers given to the Royal Society of Edinburgh, Kelvin tested **uranium** rays as well as X-rays and found that both caused the air to become electrified but did not explain how it happened. **Marie Curie** used his experiments as a springboard for her own research but explained the phenomena by demonstrating that one element could be transformed into another, an idea that Kelvin thought smacked of alchemy. Although Kelvin was solicitous toward Marie after Pierre's death, he continued to disapprove of her conclusion that **radium** was a new element. On 9 August 1906, in the correspondence columns of the *Times*, he launched a broadside against her ideas. Although Marie did not doubt that she was correct, she did not attack Kelvin himself. After Kelvin's death in 1907, the opposition to radium as a new element evaporated.

KIRCHOFF, GUSTAV ROBERT (1824–1887). Kirchhoff and **Robert Bunsen** developed a method for using spectroscopy to identify elements in the 1860s.

KLAPROTH, MARTIN HEINRICH (1743–1817). The 18th-century chemist Martin Klaproth discovered **uranium**, zirconium, and cerium and named titanium and tellurium. He extracted uranium, a gray metallic substance, from the gooey black compound known as **pitchblende** from the **Joachimsthal region** on the German–Czech border. He named this element uranium after the planet Uranus, discovered by the astronomer William Herschel in 1781. The Curies later isolated two new elements, **radium** and **polonium**, from the uranium-rich pitchblende.

L

LABOUISSE, EVE DENISE CURIE (1904–2007). The younger daughter of **Marie** and **Pierre Curie** was the only member of the Curie family who was not a scientist. Her interests were in the humanities, especially writing and the arts. She scarcely knew her father because he was killed when she was two years old. Although Eve as a child sometimes felt neglected by her busy mother, she later appreciated her and, after Marie's death, wrote her biography (*Madame Curie: A Biography by Eve Curie*). The English edition was translated by Vincent Sheean and published in 1938. Eve carefully researched available documents and letters and even went to **Poland** to gather information about Marie's family, childhood, and youth. This book was popular worldwide and was simultaneously published in many countries, including in the United States, where it won the third annual National Book Award for nonfiction. Eve's personality was very different from both her serious and often dour mother and her older sister Irène. She was vibrant and fun loving. Both Curie girls graduated from a private high school in Paris, the College Sévigné. Eve Curie had a fascinating life on her own. She became a very competent pianist and performed on stage many times. During the Second World War, she left Paris after the German invasion and went to England, where she joined the Free French Forces of Charles de Gaulle and actively opposed the Nazis and the Vichy government in France. To punish her, this government deprived her of her French nationality and confiscated her property. While she was in England during the war years, she met Winston Churchill.

On a trip to the United States in 1940, Eve met Eleanor Roosevelt at the White House and wrote articles for American newspapers. After returning to Britain, she became a war correspondent and traveled to Africa, the Soviet Union, and Asia in 1941 and 1942. Her reports from this extensive journey were published in American newspapers and were collected in a book titled *Journey among Warriors*, which was nominated for the Pulitzer Prize. In her articles and books, she reported on interviews with people from all walks of life from the shah of Iran, the leader of nationalist China Chiang Kai-shek, and Mahatma Gandhi to soldiers, factory workers, political leaders, and scientists. After her year as a foreign correspondent, she volunteered for the Free French women's medical corps for the Italian campaign and was promoted to lieutenant in the French First Armored Division. She took part in the landing with French troops in Provence in 1944 and was awarded the Croix de Guerre for her wartime service.

After the war, she served as a coeditor for a daily newspaper for five years while taking an active political role in de Gaulle's government. She was active in the early years of the United Nations and appealed to that organization for recognition of the state of Israel. From 1952 to 1954, she was a special advisor to the first secretary general of NATO. In 1954, Curie married Henry Richardson Labouisse Jr., an American politician and diplomat. Eve became a United States citizen in 1958. Labouisse became the executive director of UNICEF from 1965 until 1979, and he and Eve (who also worked for UNICEF) visited many Third World countries. Labouisse received the Nobel Peace Prize in 1965. After Labouisse died in 1987, Eve lived in New York City. On her 100th birthday in 2004, she had a visit from Kofi Annan, secretary general of the United Nations, and she received congratulatory letters from President George W. Bush and French President Jacques Chirac. Although her accomplishments were spectacular, she always noted that in a family that had five Nobel Prizes (two for Marie, one for Pierre, one for Iréne and Fédéric Joliot Curie, and one for Henry Labouisse), she was the unaccomplished one. However, when her life is examined, her accomplishments in different areas were spectacular. She died in her sleep on 22 October 2007 at the age of 102.

LAMOTTE (MONSIEUR). Before meeting **Pierre Curie**, Marie Sklodowska had a suitor, a Monsieur Lamotte, about whom little is known, although he courted her and hoped the relationship would become reciprocal. In 1894, as Marie took her last series of examinations at the **Sorbonne**, he knew she planned to return to **Warsaw** and live with her father and make a living by teaching. His farewell letter is all that remains. He thanked Marie for being his friend and, while he realized that they would never meet again, pledged that he would not forget her.

LANGEVIN, EMMA JEANNE DEFOSSES (1874–1970). Jeanne Desfosses Langevin was the wife of **Paul Langevin** and the mother of Jean Langevin, Àndre Jacques Langevin, Madeleine Varloteau, and Hélène Henriette Langevin. Her marriage to Paul Langevin was in trouble from the very beginning, and Paul knew that he had made a mistake. Jeanne was even violent toward Paul, and on one occasion hit him with an iron chair. However, a period of relative calm ensued, and the couple, at least on the surface, appeared to be getting along. It seemed that Paul merely tolerated Jeanne because he loved their children and was concerned that a divorce would hurt them. It is probable that both parties were to blame, with Paul being autocratic and Jeanne unwilling to accept his orders. Paul was a good friend and colleague of **Pierre Curie** and mourned his loss.

After Pierre's death, Paul began to spend more time with **Marie Curie**. Marie's admirers had the image of her as a grieving widow who had done exciting work with **radium**, but this image was destroyed when a group of letters from Marie Curie to Paul Langevin were stolen. The newspapers eventually got hold of these letters but were quiet for eight months. After a particularly acrimonious argument—Jeanne claimed Paul had hit her in the face for cooking badly, and he countered that she had hurled insults at him in front of the children—he took the two boys without her permission and thus opened himself to a lawsuit. Jeanne's revenge by exposing the letters was almost successful. However, many people dismissed them as the rantings of a scorned wife. Those who believed Jeanne mounted a hate campaign against Marie, and her reputation was sullied and her health compromised. The *New York Times* reported on 9 December 1911 that the case against Paul Langevin had been dismissed. Jeanne filed for divorce and was granted custody of their four children. *See also* CURIE–LANGEVIN AFFAIR; DUELS INVOLVING M. CURIE.

LANGEVIN, PAUL (1872–1946). Paul Langevin was a French physicist who had an outstanding career and was known among other things for his work on paramagnetism and diamagnetism as well as on the Langevin equation. He was awarded the Hughes Medal (1914) and the Copley Medal (1940). Langevin was elected to the **Académie des Sciences** (French Academy of Science) (1934) and was elected a fellow of the **Royal Society**. During World War I, he put his theoretical knowledge into practical use by applying **Pierre Curie**'s **piezoelectric** effect to detect submarines through echolocation. Politically active, he was one of the founders of an antifascist organization in 1934 and president of the Human Rights League. He also joined the French Communist Party. His opposition to fascism resulted in his arrest, and he was held under house arrest by the Vichy government for most of World War II. Langevin was a student of Pierre Curie at the **Sorbonne** and received his Ph.D. degree from that institution. Previously he had studied with **J. J. Thomson** at the Cavendish Laboratory at Cambridge University in 1902. He married Emma Jeanne Desfosses, and the couple had four children: Jean, Ándre, Madeleine, and Hélène. His private life became public after his supposed affair with **Marie Curie** after the death of Pierre. *See also* LANGEVIN, EMMA JEANNE DEFOSSES; CURIE–LANGEVIN AFFAIR.

LANGEVIN-JOLIOT, HÉLÈNE (b. 1927). Hélène Langevin-Joliot was the daughter of **Irène Joliot-Curie** and **Frédérick Joliot-Curie**. Her maternal grandparents were **Marie** and **Pierre Curie**. Her husband, Michel Langevin, was the grandson of the physicist **Paul Langevin**, who had an affair with the widowed Marie Curie. She became a nuclear physicist herself and in 2012 became a professor of nuclear physics at the Institute of Nuclear Physics at the University of Paris. She is known for encouraging **women** to pursue careers in scientific fields. She has a son, Ives (b. 1951), who is an astrophysicist. *See also* CURIE–LANGEVIN AFFAIR.

LANKESTER, E. RAY (1847–1929). The son of **H. Edwin Lankester**, Ray Lankester was a comparative anatomist who studied Protozoa, Mollusca, and Arthropoda. He was knighted in 1907 and awarded the Copley Medal of the **Royal Society** in 1913. When the Curies visited London in 1903, they met Ray Lankester, who like his father was one of the elite British scientists. He assumed the Linacre chair at Oxford in 1891.

LANKESTER, H. EDWIN (1814–1874). Edwin Lankester was one of the elite of British intellectual society. He was a medical doctor and president of the British Association for 25 years. He also was a fellow of the **Royal Society**, president of the Royal Microscopical Society, and founded the *Quarterly Journal of Microscopical Science*. Although he did microscopic research and wrote articles on botany, some of them popular accounts, he continued to practice medicine. He met **Marie** and **Pierre Curie** when they went to London in 1903. *See also* LANKESTER, E. RAY.

LE CHÂTELIER, HENRI (1850–1936). Henri Le Châtelier was a French chemist who is known for Le Châtelier's principle predicting the effect a changing condition has on a system in chemical equilibrium. He discovered that when a chemical system is subjected to a change (such as a change in concentration, temperature, or pressure), the system always acts to oppose these disruptions and tries to restore equilibrium. When **Marie Curie** was working on her early research project in 1897 involving the way the magnetic properties of various tempered steels varied with their chemical composition, Le Châtelier was one of her advisors. *See also* EARLY RESEARCH OF M. CURIE.

LEBON, GUSTAVE (1841–1931). After **Wilhelm Röntgen** discovered X-rays, the search for other forms of radiation proliferated. LeBon called his new radiation "**black light**," but it was not given much credence. However, **René Blondlot** had better luck with his **N rays**.

LENARD, PHILIPP (1862–1947). Philipp Lenard (Philipp Eduard Anton von Lenard) was a German physicist who won the Nobel Prize for Physics in 1905 for his research on cathode rays. It was his work on these rays that **Wilhelm Röntgen** tried to repeat when he finally discovered X-rays.

LESLIE, MAY SYBIL (BURR) (ca. 1887–1937). May Sybil Leslie graduated with first-class honors in chemistry from the University of Leeds. After she was awarded the 1851 Exhibition Science scholarship for 1909–1910, she went to Paris to work in **Marie Curie**'s laboratory. Her assignment was to search for new **radioactive** elements in the mineral thorite. Unsuccessful in this search, she then investigated **thorium** and its decay products. She used diffusion to find the molecular weight of thorium emanation (later known as **radon**). Her scholarship was continued for a second year. In 1911–1912, she continued her work on radioactive gases in **Ernest Rutherford**'s laboratory in Manchester. After leaving Rutherford's laboratory, she taught at University College of North Wales, where she installed a physical chemistry laboratory and organized a course in physical chemistry for honors students. In 1915, she joined the war effort as an industrial research chemist. In 1918, she received a doctorate of science for her research, especially on radioactive substances and the production of explosives and use of their by-products. She returned to Leeds and spent the rest of her career there. She married, but her husband died several years after their marriage. *See also* WOMEN SCIENTISTS IN M. CURIE'S LABORATORY.

LEUKEMIA. Most kinds of leukemia can be caused by past radiation exposure. Early examples of this correlation surfaced when the **radium**-dial painters provided proof that ingested radium, like calcium, could be lodged in the bones, cause damage to blood-forming tissues, and cause abnormalities, including anemia and leukemia.

LIPPMANN, GABRIEL (1845–1921). Jonas Ferdinand Gabriel Lippmann was born in Luxembourg. The family moved to Paris in 1848. First tutored by his mother, he then attended the Lycée Napoléon, followed by the École Normale Supérieure. After failing the aggregation examination, which would have allowed him to become a teacher, he studied his preferred subject, physics. The French government sent him to Heidelberg University, where he received a doctorate, summa cum laude, in 1874. He became knowledgeable in the German scientific education system, which was superior to the French system. He returned to Paris in 1875, where he studied until he became professor of physics at the **Sorbonne**. He won the Nobel Prize for Physics in 1908. Lippmann allowed **Marie Curie** to work in his laboratory at the Sorbonne, although she complained about the small space he provided. As did the Curie brothers, Lippmann worked in **piezoelectricity**. Lippmann also sought funds for Marie Curie's research that resulted in her doctorate. When Curie was first mentioned as a possible candidate for the Nobel Prize, there was an effort to omit her name. Lippmann was one of the four members of the **Académie des Sciences** who wrote a nominating letter for the prize and ignored Curie's accomplishments even though he had been an earlier mentor. Later, he reviewed her accomplishments in her unsuccessful attempt to become a member of the Académie des Sciences.

LODGE, SIR OLIVER (1851–1940). Sir Oliver Lodge was an industrialist who became interested in electricity and electromagnetic waves and became a part of the British scientific community. When the Curies were in England during 1903, Lodge was one of the scientists they met. He also joined other 19th-century scientists in studying psychical phenomena, especially telepathy. **Pierre Curie** dabbled in this area immediately before his death.

M

"**MADAME SKLODOWSKA.**" After **Marie Curie** accepted her **Nobel Prize** in Stockholm, she became very ill. Her close friends thought that the illness was precipitated by the fallout from the **Paul Langevin** scandal. She was hospitalized and diagnosed with a kidney and ureter infection caused by old lesions. After the acute symptoms subsided, she returned to work in her laboratory. In late March 1912, she had surgery to remove the lesions and was very ill for a long time. Before the surgery, she had weighed about 123 pounds, but afterward her weight dropped to 103 pounds. Although her physical problems were very real, she was haunted by the thought that she had disgraced the Curie name. In order not to besmirch the family, she insisted on being called "Madame Sklodowska." Her family and close friends rallied around her. Sister Bronia rented a small house for her outside Paris, and **Jacques Curie** was very supportive of his former sister-in-law. She had a relapse in June and spent some time in a sanatorium in the mountains of Savoy. When she was strong enough, she visited her English scientist friend **Hertha Ayrton**. In England, she gradually increased her strength and was sufficiently strong enough to return to Paris. In December 1912, she began her experimental work again. To daughter Irène's delight, she became Madame Curie again. *See also* CURIE–LANGEVIN AFFAIR; HEALTH PROBLEMS OF M. CURIE.

MARCONI, GUGLIELMO (1851–1940). The inventor and electrical engineer Marconi is known for his work in pioneering long-distance radio transmission, Marconi's law, and a radio-telegraph system. He is credited with the invention of radio. He shared the 1909 Nobel Prize in Physics with Karl Ferdinand Braun for their contributions to the development of wireless telegraphy. Many scientists thought that **Edouard Branley** should also have shared the prize in physics. When **Marie Curie** became a candidate for the French Academy of Sciences (**Académie des Sciences**), her opponents were hunting for a candidate who could possibly defeat her. They found one in Branley, who had been defeated as a candidate two times previously. Many French academics felt that he deserved the Nobel Prize in 1909 when Marconi and Braun had received it. After a close election, Branley defeated Curie.

MARIĆ, MILEVA (1875–1948). Marić was a Serbian mathematician who was **Albert Einstein**'s first wife and the mother of his children. *See* EINSTEIN, MILEVA MARIĆ.

MARRIAGE OF IRÈNE CURIE TO FRÉDÉRIC JOLIOT. In 1926, Irène married chemical engineer **Frédéric Joliot**. When they married, they hyphenated their surnames to Joliot-Curie. The couple had two children, Hélène, who also became a noted nuclear physicist at the University of Paris, and a son, Pierre, who became a biochemist at Centre National de la Recherche Scientifique. Just as **Marie** and **Pierre Curie** collaborated on their research, so did Irène and Frédéric Joliot-Curie. This collaboration resulted in the **Nobel Prize for Chemistry** in 1935.

MARSDEN, ERNEST (1889–1970). His experiment with **Hans Geiger** (Geiger–Marsden experiment) suggested by **Ernest Rutherford**, led to the discovery of a new theory on the structure of the atom, which led Rutherford to the Nobel Prize in Chemistry.

MARXISM. Originating from the work of the 19th-century philosophers Karl Marx (1818–1883) and Friedrich Engels (1820–1895), Marxism uses a materialistic interpretation of historical development and a dialectical vision of social transformation. When Maria Sklodowska was a young woman, this movement had followers in **Poland**. The **Floating University** she and her friends were a part of was fascinated by ideas that they hoped would solve Poland's problems. As opposed to **positivism**, another philosophy gaining credence at that time, Marxism was more radical, rejecting collaboration with occupying powers and supporting revolutionary change. Maria Sklodowska preferred the positivists' ideas of gradual change with scientific solutions to Poland's problems.

MAUNDER, ANNIE SCOTT DILL RUSSELL (1868–1947). Annie Russell Maunder exemplifies the importance of collaboration between husband and wife in the history of science. The Curies, of course, are the prime example of this kind of partnership, but they certainly were not the only ones. Irish English astronomer Annie Russell, born in Northern Ireland and educated at **Girton College, Cambridge**, collaborated with her husband, **Walter Maunder**. This relationship illustrates the importance of collaboration in making science an acceptable occupation for **women**. Before her marriage, Russell, who had won the highest mathematical award open to women, the Senior Optime in the Mathematical Tripos, was employed as a low-paid "lady computer" in the **Royal Greenwich Observatory**. Here she met Walter Maunder, the head of the Solar Photography Department and founder of the British Astronomical Association (BAA). This organization welcomed female members. Although she worked in collaboration with her husband, Annie Maunder worked independently of him as well. Her work in astronomy included a photographic survey of the Milky Way. Her favorite research subject was the sun. She theorized that the earth influences the numbers and areas of sunspots and sunspot frequency from the eastern to the western edge of the sun's disk as viewed from the earth. She also postulated that changes in the sun trigger climate changes on earth.

MAUNDER, WALTER (1851–1928). Walter Maunder, the spouse of **Annie Maunder**, is best remembered for his study of sunspots and the solar magnetic cycle known as the Maunder minimum. He also did many observations on Mars, concluding that the so-called canals were optical illusions. He was important in the formation of the British Astronomical Association (BAA), an organization that was open to all classes of society and especially **women**. A member of the elitist Royal Astronomical Society himself, he saw the importance of forming a society open to everyone interested in astronomy. After the death of his first wife, Edith, he married Annie Scott Dill Russell, and they collaborated throughout their lives, illustrating the importance of spousal collaboration in the history of science.

MELONEY, MARIE MATTINGLY (MISSY) (1878–1943). Missy Meloney was the journalist who in 1920 was editor of *The Delineator*, a woman's magazine, and managed to interview a skeptical **Marie Curie**, who had turned away many journalists after her unpleasant experience with them following the **Curie–Langevin affair**. Meloney managed to write a letter to a skeptical Curie that resulted in a short interview. The experienced interviewer Meloney confessed that she was unusually timid when the two talked and that it was Curie who tried to put her at ease. During this interview, they talked about the scarcity of **radium** in France. During that time, a gram of radium cost $100,000. Surprisingly, the two **women** became friends. Meloney resolved to convince 10 wealthy American women to each contribute $10,000 to buy a gram of radium for Curie. Meloney convinced Curie that she did not need to fear that the American press would excoriate her for the Curie–Langevin affair. Meloney made sure that the press would cooperate. On 7 May 1921, Marie Curie and her two daughters, Irène and Eve, traveled to the United States to accept the gram of radium. Meloney was especially interested in radium as a cure for cancer. Although Marie Curie was unaware of it, Meloney had undergone radium treatment for cancer after Curie's return from her first trip to the United States. Since the doctors had informed Meloney that the treatment was experimental, she did not inform Curie. However, their mutual friend **Loie Fuller** did so. Meloney had gone against the advice of most of the surgeons, who advised her to have a mastectomy, and accepted that of the one who insisted that radium needles would give her an excellent chance of survival. This treatment seems to have been successful, as she lived until 1943, when she died of influenza. See also UNITED STATES, M. CURIE'S 1921 TRIP TO THE.

MEYER, STEFAN (1872–1949). Meyer was an Austrian physicist who worked in the field of **radioactivity**. He became acting director of the Institute for Radium Research in Vienna that opened in 1910. In the spring of 1910, **Ernest Rutherford** communicated to **Marie Curie** the importance of developing an international standard for **radium**. He asked Curie to send a sample of radium from her laboratory so that he could compare it with his own radium. They agreed that a standard was needed in order to guarantee agreement between results found from different laboratories, to ensure accuracy in medical applications, and to be sure of the reliability in the manufacture of radium. Curie agreed to prepare the standard but then insisted that she must retain it in her laboratory. Rutherford insisted that an international standard must not be in the hands of a private person and explained the dilemma to Meyer, who was secretary of the International Radium Standard Committee. Many on the committee agreed that a duplicate placed in one of the French bureaus would alleviate the problem. However, although Curie eventually agreed to this solution, there were other problems to be addressed. Curie wanted the radium from her laboratory to be replaced, and some committee members, especially Stefan Meyer, questioned Curie's results. Meyer prepared his own standard and brought it to Paris for comparison. To everyone's relief, when the two samples were compared, they were in essential agreement. See also RADIUM, INTERNATIONAL STANDARD FOR.

MICHALOWSKA, HENRIETTA. Henrietta Michalowska was **Maria Sklodowska Curie**'s cousin, with whom she corresponded frequently as a young woman. During the four years she worked at Szczuki as a **governess**, Maria often confided in Henrietta about her feelings. When Maria was rejected by **Kazimierz Zorawski**, she expressed her disappointment and bitterness to Henrietta.

MILLIKAN, ROBERT (1868–1953). American physicist Robert Millikan won the Nobel Prize in Physics in 1923 along with many other awards. He is known, among other discoveries, for the oil-drop experiment, in which he measured the charge of a single **electron**. After **Marie Curie**'s death, he characterized her work on radiation and the radiation emitted by it in an obituary in the *New York Times* as a starting point for the new developments in physics. He contended that her ideas convinced scientists that the heavens are not eternal and changeless, for atomic transformations take place in nature constantly. He also mentioned her role in establishing world peace through her membership on important committees of the League of Nations.

MITTAG-LEFFLER, GÖSTA (1846–1927). Gösta Mittag-Leffler was a Swedish mathematician who was an advocate for **women**'s rights. In 1903, the members of the Swedish Academy of Science were prepared to award the **Nobel Prize for Physics** jointly to **Pierre Curie** and **Henri Becquerel**, leaving out **Marie Curie**. Mittag-Leffler was a member of the prize committee and wrote to Pierre explaining that he and Becquerel were to be nominated. Pierre wrote back saying that he would be honored to be considered for the prize but insisted that Marie should be nominated also.

Mittlag-Leffler was instrumental in finding a way to include Marie Curie.

MORTIER, PIERRE. Mortier was a writer for *Gil Blas* and fought a **duel** in 1911 with **Gustav Téry**, who had written a column in his newspaper *l'Oeuvre*. Mortier was wounded in the arm defending **Marie Curie**'s honor. *See also* CURIE–LANGEVIN AFFAIR.

MOTHER, M. CURIE AS A. Marie Curie loved her two children and was intensely concerned about being a proper mother, although she was not, by nature, a warm person. When Irène was born, Marie was determined to continue her research but was also solicitous about caring for the baby. Although she gave up nursing Irène and acquired a wet nurse because she was not thriving, Marie continued to bathe, change, and dress her. While Marie worked in the laboratory, Irène was cared for by a nurse. Whenever Irène had any physical problems normal for a young child, such as teething, colds, or minor accidents, Marie was extremely concerned. She often was troubled that she was not doing the right thing in caring for Irène and asked for advice when she was perplexed as to what to do. As Irène got older, her grandfather **Eugène Curie** became her major caregiver. In her diary, Marie described Irène's behavior and physical changes in a school notebook as she would do in her laboratory book. She recorded her weight every day, her diet, when she got her first tooth, and general accomplishments. Daughter Eve in her biography of her mother described one of these entries:

> Irène has cut her seventh tooth, on the lower left. She can stand for half a minute alone. For the past three days we have bathed her in the river. She cries, but today (fourth bath) she stopped crying and played with her hands in the water. She plays with the cat and chases him with war cries. She is not afraid of strangers any more. She sings a great deal. She gets up on the table when the baby is in her chair. (Eve Curie, *Madame Curie*, 163)

When Eve was born in 1904, her father was only a part of her life for two years. Marie was busy trying to control her own emotions after **Pierre Curie**'s death and often was overwhelmed with the responsibility of raising her daughters without a father. Eugène Curie, their grandfather, as well as a governess removed part of the strain from Marie. Although clearly Marie was a doting mother, her personality did not allow her to be demonstrative, and it may have influenced the way the children's personalities evolved. Irène's personality and interests were similar to her mother's, so each had an implicit understanding of the other, but Eve had very different interests. Eve later confessed that she had not had sufficient attention from her mother as a young child, but as a young adult she developed a strong emotional tie to her. After Pierre's death, Marie, the girls, and their grandfather Eugène moved to a western suburb, Sceaux. Sceaux, though it was farther from the **Sorbonne**, had advantages. It was Pierre's youthful home and close to his grave. It also had room for a garden and a place the children could play outside.

MOUTON, HENRI (1869–1935). Mouton was a biologist at the **Pasteur Institute** and taught natural science in **Marie Curie**'s homeschool project for **Irène Curie** and other children of scientists.

N

N RAYS. After the discovery of **X-rays**, many scientists joined the search of another hitherto unknown type of radiation. **René Blondlot** claimed that he had found a new radiation and called these rays N rays after his hometown, Nancy.

NEUMANN, ELSA (1872–1902). **Marie Curie** was determined to obtain a doctoral degree, but during the time she was a student at the **Sorbonne** no woman in Europe had completed this degree. Elsa Neumann, a German physicist, was working on a doctorate at that time but, unlike Curie, was unmarried and did not have family responsibilities. Neumann attended the University of Berlin and then studied physics at the Max Planck Institute, where she eventually completed the degree. In order to collect her doctorate, Neumann had to receive special permission from the Ministry of Education. Permission was granted on 18 February 1899, making her the first woman physicist to receive a Ph.D. degree. The difficulties that Curie, as a wife and mother, had to overcome seemed insurmountable. She also faced the disapproval of her colleagues, many of whom were convinced that no married woman and especially one with children could earn a doctorate. *See also* WOMEN.

NEW YORK TIMES. The *New York Times*, based in New York City, was founded in 1851 and is the winner of 127 Pulitzer Prizes. It weighed in on the **Paul Langevin** scandal, reported on the lives and works of both **Marie** and **Pierre Curie**, and produced numerous articles on the **radium**-fundraising campaign. On the Langevin affair, the *Times*' conclusions were very different from those in *l'Oeuvre*. The *Times* editorialized that it would be impossible to tarnish the achievements of a brilliant woman who "has been made, rather late in life, the heroine of a somewhat scandalous romance." This newspaper blamed **Jeanne Langevin** but did not judge whether or not the charges were justified. However, it indicated that "there are hints of deliberate mischief-making in the case," that the letters may not be genuine, and even if they were true they had nothing to do with Curie's importance as a scientist ("Editors in Duel over Mme. Curie," *New York Times*, 5 February 1910). The newspaper continued to print detailed articles on circumstances in Curie's life and scientific discoveries. *See also* CURIE-LANGEVIN AFFAIR.

NOBEL PRIZE FOR CHEMISTRY, 1911, M. CURIE. Four days after the Parisian newspaper *Le Journal* published an article that purported to prove on the basis of stolen letters that **Marie Curie** and **Paul Langevin** had been involved in an adulterous affair, Marie received a telegram informing her that she had been awarded the Nobel Prize in Chemistry. The prize was awarded for discovery of the new elements **radium** and **polonium** by the isolation of radium, and the study of the nature and compounds of radium. The Nobel Committee was concerned about scandal after the appearance of the newspaper article but was mollified after Marie's denial as well as that of the man who wrote the article. She not only was the first woman to receive the Nobel Prize but one of only a few (and the only woman) to receive two prizes. *See also* CURIE–LANGEVIN AFFAIR; WOMEN.

NOBEL PRIZE FOR CHEMISTRY, 1935, I. JOLIOT-CURIE AND F. JOLIOT-CURIE.

In 1931, Irène and Frédéric began their serious collaboration and combined their separate research projects on the study of atomic nuclei. The Joliot-Curies joined other laboratories involved in studies of the atom, some of which glimpsed the important applications of this work. Although their experiments with gamma rays identified the positron and neutron, they did not recognize the significance of their results. In 1934, the couple performed an experiment that was to earn them the Nobel Prize. They irradiated the nucleus of the aluminum atom with **alpha particles**. When they stopped the bombardment, the substance produced was radioactive and the newly discovered positrons were given off. They realized that they had converted the original metal into a radioactive isotope of silicon. They had discovered artificial **radioactivity**. Like the alchemists of old, they could change one metal into another. This discovery resulted in the Nobel Prize in Chemistry for the Joliot-Curies in 1935.

NOBEL PRIZE FOR PHYSICS, 1903, M. AND P. CURIE AND H. BECQUEREL.

In December 1903, **Marie** and **Pierre Curie** and **Henri Becquerel** were jointly awarded the Nobel Prize for Physics. Becquerel won half of the prize for his discovery of spontaneous **radioactivity**, and Pierre and Marie shared the other half, each receiving a fourth of the prize for their joint researches on the radiation phenomena discovered by Becquerel. Becquerel went to Stockholm to receive his award, but the Curies, both of whom were unwell, were absent, pleading uninterruptible teaching schedules for their absence. In actuality, Marie had lost a child prematurely after one of their bicycle rides. She had been exposed to large doses of radiation during this pregnancy. Not until June 1905 did the Curies travel to Sweden, where Pierre gave the lecture required of Nobel laureates. Marie Curie became the first **woman** to win a Nobel Prize.

Plaque announcing the winners of the 1903 Nobel Prize in Physics, shared between Marie and Pierre Curie and Henri Becquerel

L'OEUVRE. This weekly newspaper was founded, edited, and largely written by **Gustave Téry** in 1909. Téry had been a journalist for a feminist newspaper, *La Fronde*, but by the time he founded *l'Oeuvre* his views had veered to the right after he was rejected by academia. He became rabidly anti-Semitic and antiacademic. He used the **Curie–Langevin affair** as an opportunity to discredit academia. On 23 November 1911, *l'Oeuvre* published excerpts from letters between **Marie Curie** and **Paul Langevin**. The French public was incensed by the letters and took the side of **Jeanne Langevin**, a respectable French woman, against Curie, a dubious foreign woman.

ORZESZKOWA, ELIZA (1841–1910). Eliza Orzeskowa was a Polish positivist novelist who became a part of Maria Sklowdowska Curie's intellectual persona for the rest of her life. Orzeszkowa not only wrote novels but by espousing positivism tried to educate her readers in order to change their attitudes, hoping to eliminate class, race, and gender prejudices. She ignored the part of **Auguste Comte**'s **positivism** that considered **women** inferior by nature. She espoused the education of the masses, the development of science, and class discrimination and was a strong proponent of evolution and agnosticism. All of these ideas influenced the young Maria Sklodowska.

P

PANTHEON OF PARIS. This neoclassical building was begun under King Louis XV in 1758 and completed in 1765. This structure houses, among other treasures, the 67-meter (220-foot) Foucault pendulum. Many of the great people of France, including Victor Hugo, Marcellin Berthelot, Rousseau, Voltaire, and Emile Zola, are buried in its necropolis. Although **Marie** and **Pierre Curie** were originally buried in Sceaux, they were moved on 20 April 1995 to the Pantheon. Marie Curie was the first woman to be buried for her own accomplishments in this national monument.

PASTEUR INSTITUTE (INSTITUT PASTEUR). The Pasteur Institute was founded in 1887 by the French chemist and microbiologist Louis Pasteur. From its inception, Pasteur brought together scientists from different specialties, and his successors continued this tradition. **Marie Curie** was committed to forming an institute devoted to **radioactivity** in honor of **Pierre Curie**. It was to be jointly established by the **Sorbonne** and the Pasteur Institute. In order for the radiology institute to become a reality, the two entities had to agree on funding and the responsibilities of each for the implementation of policies. **Émile Roux**, the physician director of the Pasteur Institute, was an advocate for the radiology institute. He agreed that the Pasteur Institute would finance the building of two separate laboratories, one to be directed by Marie Curie and funded by the Sorbonne, and the other would be directed by the medical researcher Dr. **Claudius Regaud**, who would study the medical applications of radioactivity. The two buildings would be located next door to each other, and together they would be known as the Institut du Radium. *See also* RADIUM INSTITUTE.

PEREY, MARGUERITE (1909–1975). Marguerite Perey was a French radiochemist who originally was trained as a chemical technician at the École d'Enseignement Technique Feminine, where she received a diploma in 1929. After graduation, she worked under **Marie Curie** at the **Radium Institute**, where Curie served as her mentor, making Perey her personal assistant and helping her develop her chemical knowledge and skills. In 1939, Perey discovered element 87 and named it francium. Francium was the last of the missing elements existing in nature; the rest had to be produced artificially. She completed her studies at the **Sorbonne** and eventually received a doctorate in physics in 1946. After she was appointed to a new chair of nuclear chemistry at the University of Strasbourg, she developed a program in radiochemistry. She received many honors for the discovery of francium, including election to the **Académie des Sciences**, where she became its first female corresponding member. Perey's successful career demonstrates Marie Curie's ability to teach others, often women, to be outstanding scientists. As did too many of the early workers in radiation, Perey died from cancer. *See also* WOMEN.

PERRIN, ALINE (1899–1991). Aline Perrin was the daughter of **Henriette** and **Jean Perrin** and the sister of **Francis Perrin**. She was a childhood friend of **Irène Curie**. During the **Curie–Langevin affair**, Irène stayed with the Perrins. Although she was reluctant to leave her mother, she finally agreed because of her friendship with Aline.

PERRIN, FRANCIS (1901–1992). Francis Perrin was the son of **Henriette** and **Jean Perrin** and the brother of **Aline Perrin**. Like his father, he became a physicist. He obtained a doctorate in mathematical sciences for his thesis on Brownian motion. From 1946 to 1972, he was professor at the College de France and held the chair in atomic and molecular physics.

PERRIN, HENRIETTE JEANNE EUGENIE BLANCHE DUPORTAL (1869–1938). Henriette was the wife of physicist **Jean Perrin** and the mother of **Aline** and **Francis Perrin**. Not only was she the Curies' next-door neighbor, she was a good friend of both Curies and was especially supportive of Marie after Pierre's death. After hearing the news of Pierre's death, Marie asked Henriette to take care of Irène while she processed the tragedy.

PERRIN, JEAN (1870–1942). Physicist Jean Perrin had a long history with the Curies. When **Pierre Curie** had applied for a vacant position in the faculty of the **Sorbonne**, Perrin, a younger physicist who had attended all of the prestigious schools, was appointed over a disappointed Pierre. This event did not disrupt the friendship of the two physicists. Jean Perrin was married to **Henriette Jeanne Eugenie Blanche Duportal** and was the father of **Aline Perrin** and **Francis Perrin**. The Perrins and the Curies were next-door neighbors. After **Albert Einstein** published his theoretical explanation of Brownian motion, Perrin did the experimental work that tested and verified Einstein's predictions. He received the Nobel Prize in Physics in 1926 for this work and other work on the discontinuous structure of matter. He taught the children chemistry under **Marie Curie**'s plan for a cooperative school. On 19 April 1906, Perrin and Curie were together at a meeting and wended their way back to the Latin Quarter afterward. They split up, and Pierre continued alone toward the Institut de France, but on the way he was killed in an accident. Perrin and the dean of the faculty of the Sorbonne, **Paul Appell**, were the ones who informed Dr. **Eugène Curie** of his death, and when Marie returned, Appell reviewed the facts again. After Pierre's death, the Perrins were very supportive of Marie Curie. *See also* DEATH OF P. CURIE.

PHOSPHORESCENCE. Phosphorescence is a quality that makes certain substances glow in dark. It is often confused with fluorescence because both emit more light than other objects next to them. Although fluorescent molecules harness energy from an energy source, phosphorescence substances continue to emit light after the energy source is removed. The theoretical explanation was developed by **Philipp Lenard**. Lenard observed that **cathode** rays would illuminate substances some distance away from them if they were coated with a phosphorescent substance (one that emits light without an appreciable amount of heat). He developed a new explanation and published it as his theory of electron excitation and luminescence. **Wilhelm Röntgen**, when attempting to reproduce Lenard's study, discovered X-rays.

PICKERING, EDWARD (1846–1919). During the late 19th and early 20th centuries, Pickering, the director of the **Harvard College Observatory**, provided American women, educated in the new women's colleges, with paid positions at this observatory. As astronomy moved from a strictly observational field into the new field of photographic astrophysics, Pickering needed fewer observers (men's work) and more assistants (**women**'s work) to classify as cheaply as possible thousands of photographic plates that his equipment was generating. This period of time was important for women all over the world. **Marie Curie** was a partial beneficiary of this movement.

PIEZOELECTRIC QUARTZ BALANCE. This balance resulting from **Pierre** and **Jacques Curie**'s experiments on **piezoelectricity** became important in the delicate measurements that resulted in the discovery of **radium**. *See also* CRYSTALLOGRAPHY.

PIEZOELECTRICITY. This term described the early research of the Curie brothers, Pierre and Jacques. Together the brothers published nine papers on the phenomenon. They found a way of measuring the relationship between pressure and small amounts of electricity using quartz as their material. *See also* CRYSTALLOGRAPHY; PIEZOELECTRIC QUARTZ BALANCE.

PITCHBLENDE. This heavy gooey black compound is one of the chief mineral ores of **uranium** and contains 50–80 percent of that element. It is also referred to as uraninite. The pitchblende that **Marie Curie** used was found in the **Joachimsthal region** on the German–Czech border. When Curie was working on her doctoral project, she realized that the activity of the uranium compound depended solely on the amount of uranium present and concluded that the emission of the rays described by **Antoine-Henri Becquerel** is an atomic property of the uranium and not an ordinary chemical reaction. Pitchblende was the source of the uranium she needed for her research. She determined that pitchblende was four times as active as uranium and concluded that pitchblende must contain something much more radioactive than uranium. In order to demonstrate this hypothesis, she realized that she would have to isolate the highly radioactive substance in the pitchblende. This led to her discovery of two new elements, **polonium** and **radium**.

PLUM PUDDING MODEL OF ATOM. *See* GEIGER, HANS.

POINCARÉ, HENRI (1854–1912). When **Marie Curie** was a student at the **Sorbonne**, the theoretical physicist and mathematician Henri Poincaré was one of her teachers. Poincaré excelled in many different areas. After **Wilhelm Röntgen** published his X-ray paper in 1896, Poincaré attempted to explain X-rays in a report to the **Académie des Sciences**. He observed that X-rays caused **phosphorescence** both on the wall of the vacuum tube and on a screen outside the tube that was coated with a phosphorescent substance. Poincaré was influential in finding a position for **Pierre Curie** on the faculty of the Sorbonne when, disappointed after he was passed over for the chair of physical chemistry, the Curies almost left Paris for a position in Geneva. *See also* POSITIONS, PROBLEMS WITH, FOR P. CURIE.

POLAND. The home country of **Marie Curie**, Poland has a checkered history from early modern times to the present. During the late Middle Ages and early modern era, Poland was ruled by the Jagiellon dynasty (1386–1572). The amount of land Poland occupied ebbed and flowed during this period. The Jagiellon dynasty brought land controlled by Lithuania into Poland's sphere of influence. In the early modern era (1569–1795) the Polish–Lithuanian Commonwealth was established bringing a period of stability and prosperity in Poland. It included modern-day Lithuania, Ukraine, Belarus, and Western Russia. From the middle of the 17th century, the power of Poland gradually declined, partially because of internal disorder making it vulnerable to outside intervention. During the 16th and 17th centuries, Poland became involved in conflicts with Russia, Sweden, and the Ottoman Empire.

On the death of the Polish king, Augustus II, Empress Catherine II (Catherine the Great) of Russia saw an opportunity to extend Russian influence in Poland by supporting the pliable Stanislas Poniatowski (who was her ex-lover) for king. She joined with Frederick II (Frederick the Great) of Prussia to secure Stanislas's election. In 1764, Stanislas became the king of Poland. If the election had not gone her way, Catherine was prepared to act against Poland militarily. Although Stanislas recognized that he must implement reforms in order to save his country, it was necessary for him to remain in a political relationship with Russia. His subjects were unhappy and rebelled against both the king and his Russian sponsors. Catherine eventually gained the puppet state that she desired, but the Poles revolted. This civil war merged into a foreign war that resulted in the first partition of Poland.

Even though Russia and Prussia were uneasy allies, they united because of their common enemy, Austria, which, under the rule of Joseph, was determined to take Silesia from Prussia. Thus, Prussia needed Russia's guarantee to protect Silesia from Austria. However, the geopolitical affairs became even more complex, and Russia and Prussia brought Austria into the partition. Poland's neighbors, Prussia, Russia, and Austria, agreed to partition Poland in 1772. Poland lost nearly one-third of its territory and almost one-half of its population in the first partition. Even after the first partition, in terms of territory, population, and wealth Poland was still important. Catherine feared a reborn Poland and bribed the Polish peasantry and other malcontents with money. However, even with this incentive, the Poles hated the Russians enough to conclude an alliance with Prussia.

The Polish peasantry had gained some rights after the French Revolution, strengthening and unifying Poland. The fear of a strong Poland upset the neighboring great powers. Russia invaded Poland, but although this invasion alarmed Prussia, it was afraid to interfere because its emperor, Frederick William, was beginning a campaign in France and thus was loathe to overextend. Frederick William repudiated his alliance with Poland and signed a secret agreement with Russia for the partition of his recent ally. Poland fought bravely but was deserted by its allies. The second partition gave large sections of Poland to both Russia and Prussia. Austria was left out of this partition. In what was left of Poland (less than half of her territory in 1773), the Polish people rebelled and brought on the third and final partition. This time Austria was again in on the kill. From the three partitions, Russia gained about 180,000 square miles and 6,000,000 people; Austria 45,000 square miles and 3,700,000 people; and Prussia, 57,000 square miles and 2,500,000 people. Any hope that the Polish people had of regaining control over their country through Napoleon was squelched after his defeat by Russia (1811–1812). Napoleon's Duchy of **Warsaw** was replaced by a Kingdom of Poland connected to Russia by a union with the tsar of Russia. This tsar also became king of Poland. Before Maria Sklodowska was born, there were numerous failed uprisings. The rebellious Poles were harshly defeated and gave up on the idea of liberating Poland through military means. However, they continued to rebel using civil disobedience instead. Maria Curie's parents met and married in this political climate in 1860. When Maria Sklodowska was a child in Poland, her schooling was informed by this situation.

POLONIUM. When **Marie** and **Pierre Curie** were investigating the cause of **radioactivity** in **pitchblende**, they found a product that was 330 times more active than **uranium**. This result culminated after they had used **fractional crystallization** to separate out different substances from pitchblende and found that with each fraction the crystals remaining became increasingly more radioactive. They found the element bismuth in the pitchblende refused to separate from what Marie postulated was a new element. She took a solution of bismuth nitrate that she presumed contained the new element, added hydrogen sulfide to it, collected the solid precipitate, bismuth sulfide, and found it was 150 times more active than uranium. They continued the procedure and finally came up with a product that was 300 times more active than uranium. Even though it was not pure, they were convinced that they had discovered a new element. They found that after they had eliminated both bismuth and polonium, the remaining liquid was still very radioactive. The remaining impurity was the element barium, which they knew was not radioactive. The final radioactive substance in the pitchblende was their second new element, **radium**. In July 1898, Marie announced the new substance and named it polonium in honor of Poland, her native country. During World War II, polonium was produced as part of the Manhattan Project and was a vital part of the nuclear design used in the bomb on Nagasaki in 1945.

POSITIONS, PROBLEMS WITH, FOR P. CURIE. **Pierre Curie** was a quiet, diffident person uninterested in advancement in academics. An imaginative, idealistic dreamer, he despised academic politics and refused to play the game that was required if he were to progress up the academic ladder. He believed that science was the path through which social reform could progress. Even though he claimed not to care about "advancement," he was disturbed that his abilities were not appreciated. After he obtained his *licence ès sciences* in 1878 at the **Sorbonne**, he became a laboratory assistant at the university, where he did important studies on crystals, including an important study on the phenomenon of **piezoelectricity**. Here he collaborated with his older brother, **Jacques Curie**, until Jacques married and left for a university position at Montpellier. After Jacques left, Pierre took a job at the École de Physique et Chimie Industrielles d la ville de Paris (ESPCI) (School of Industrial Physics and Chemistry) in 1882. Although he loved teaching and appreciated the freedom that he had to build a laboratory, Pierre remained at the same level without a promotion for many years. When he was recommended for a professorship, he did not apply because he did not want to go through the necessary political hassle that to do so would require. However, when he resolved to marry Marie, he decided to compromise his ideals and took a position as a technical advisor to a Parisian optical firm for 100 francs a month. He also would get a 20 percent royalty on a photographic objective lens that he had devised. Pierre presented his thesis at the Sorbonne in 1895 and invited Marie as a guest for his defense. Shortly afterward, she agreed to marry him. A professorship was created for Pierre at the École de Physique et Chimie after he presented his thesis. However, he still did not have an academic chair, although his research was well respected (**crystallography**, piezoelectricity, symmetry, and magnetism) and his joint work with Marie resulting in the discovery of **polonium** and **radium** was appreciated. When the chair of physical chemistry at the Sorbonne became available, he had every reason to think he would be appointed. In spite of his accomplishments, he was passed over and had to accept a post of assistant professor (*repetiteur*) at the École Polytechnique in order to supplement their income. Their disappointment almost caused the Curies to leave Paris for a position at the University of Geneva with a higher salary and the opportunity for Pierre to build his own laboratory. Although they were tempted, they decided to remain in Paris. Fortunately, another position opened at the Sorbonne, teaching physics to medical students. However, this position was not in the prestigious Faculté des Sciences, where Pierre wanted to be. The situation was not satisfactory, and when another position opened up in minerology he again applied and again was disappointed. Pierre simply was a bad politician and refused to flatter those in power. Another disappointment occurred when Pierre was convinced by his friends to become a candidate for membership in the prestigious **Académie des Sciences**. They assured him that his election was certain, but again he fell short. Finally, after he received the **Nobel Prize**, the French Parliament created a special professorship for Pierre at the Sorbonne. He again became a candidate for the French Academy of Sciences and, after a narrow vote, was successful. *See also* RESEARCH OF P. CURIE.

POSITIVISM. Founded by **Auguste Comte**, positivism was a philosophy that influenced the interpretation of scientific knowledge. Comte posited that certain knowledge is based on natural phenomena. This philosophical system holds that all rational assertions can be scientifically verified or are capable of logical or mathematical proof. This system consequently rejects metaphysics and theism. A study of natural phenomena, received from the senses (empirical knowledge), can lead to positive knowledge. This philosophy can be expanded to include the belief that society operates according to general absolute laws. Maria Sklodowska and her student friends in **Warsaw** stressed the social aspects of positivism, applying it to trade, science, and industrial problems in **Poland**. It was especially appealing to Marie and her sister Bronia because it stressed the importance of **women**, although Comte himself believed women were intellectually inferior. The interpretation of positivism that Marie accepted was that women, if properly educated, could contribute to reform in Poland.

Thus, many young women in Poland clustered around positivist teachers in an informal atmosphere. The ideas that Maria accepted at that time, including eliminating class, race, and gender prejudices, became a part of her intellectual persona for the rest of her life.

PREGNANCIES OF M. CURIE. Marie Curie had three pregnancies, one of which was a miscarriage. Her first pregnancy was miserable. She wrote her friend Kazia that she had continual dizziness for over two month and felt unable to work. During this pregnancy, she and Pierre were separated for the first time in July 1897, when he was with his terminally ill mother. Her father came to stay with Marie until Pierre could return. While Pierre was absent from Paris, Marie and Pierre wrote love letters to each other. Pierre sent a parcel containing baby clothes to Marie. During her eighth month of pregnancy, after Pierre returned, Marie declared that she no longer felt fatigued and was up to taking a **bicycle vacation** to Brest. Neither Marie nor Pierre seemed to think that there was anything unusual about such a trip when she was so far advanced in pregnancy. Eventually, Marie realized that she must cut the trip short, and they returned to Paris. Irène was born on 12 September 1897 and weighed 6.6 pounds. Marie first tried to nurse the baby unsuccessfully and had to employ a wet nurse. Pierre's mother had subsequently died, making it possible for Marie's father-in-law, Dr. **Eugène Curie**, to move in with the family, which made it much simpler for Marie to continue her scientific work. He adored his granddaughter and happily took over her care. Marie's second pregnancy resulted in a premature birth in 1903 after one of their bicycle rides. Born at five months, the baby girl was still living but died immediately. Marie was devastated by the loss of the baby. Although she did not make the connection, she had been exposed to extremely high doses of radiation during the pregnancy. The third pregnancy that resulted in a healthy baby Eve came after the first **Nobel Prize**. While she was pregnant, Marie gave up her teaching post at Sèvres but continued her research. Another difficult pregnancy exhausted her, but she gave birth to a healthy baby girl, Eve Denise, on 6 December 1905. She returned to her teaching job after Eve's birth although the Nobel Prize money would not have made it necessary.

PRIX LA CAZE OF THE ACADEMY OF SCIENCES. This prize of 10,000 francs was awarded to **Pierre Curie** in 1901 (or possibly 1903) and provided early support for the continued research of the Curies. The prize, created in 1875, was awarded from 1878 to 2007 by the **Académie des Sciences** for a major contribution in physics, chemistry, or physiology. It was established from the estate of doctor Louis La Caze (1798–1869).

PRZYBOROVSKA, KAZIA. Kazia Przyborovska was a girlhood friend of Maria Sklodowska. As schoolgirls at **Gymnasium Number Three**, Kazia and Maria spent many hours together and reinforced each other's dislike of Russia and love for **Poland**. When Maria left **Warsaw** for her tutoring job, they corresponded, but when Kazia wrote Maria announcing her engagement Maria was in the midst of a depression after learning that her romance with **Kazimierz Zorawski** was over. Maria was somewhat jealous of Kazia's good fortune in love, when her experience with Kazimierz had been so unfortunate.

R

RADIOACTIVITY. In a paper that **Marie** and **Pierre Curie** published in 1898, they first used the term *radioactive* to describe the behavior of **uranium**-like materials. After the Curies had used **fractional crystallization** to separate out different substances from **pitchblende**, they found that with each fraction the crystals remaining became increasingly more radioactive. However, they found that the element bismuth in the pitchblende refused to separate from what Marie postulated to be a new element. She took a solution of bismuth nitrate that she presumed contained the new element and added hydrogen sulfide to it and collected the solid precipitate bismuth sulphide and found it was 150 times more active than uranium. After many trials, they came up with a product that was 330 times more active than uranium. Even though it was still not pure, they considered it to be a new element (**polonium**). They found that after they had eliminated both bismuth and polonium, the remaining liquid was still very radioactive. The impurity that remained in the liquid was the element barium, which they knew was not radioactive. The final radioactive substance in the pitchblende was their second new element, **radium**. This element was much more radioactive than polonium and 900 times more radioactive that uranium.

RADIOACTIVITY AND CANCER. *See* RADIUM, HEALTH PROBLEMS FROM.

RADIOACTIVITY AND CATARACTS. *See* RADIUM, HEALTH PROBLEMS FROM.

RADIUM. At the end of 1898, the Curies announced the discovery of the element radium in the form of its salt, radium chloride. They published the discovery in the *Comptes Rendus* of the **Académie des Sciences**. Marie realized that although she and Pierre were convinced that radium was a new element, she must isolate it and determine its atomic weight in order to convince contemporary chemists of its elemental status. This belief on Marie's part led to her persistent efforts to purify radium, treating over a ton of **pitchblende** residues. The appalling conditions under which she worked are well known. The inadequately furnished shed that served as her laboratory leaked when it rained and froze in the winter. Her persistency paid off. In 1902, she produced one-tenth of a gram of nearly pure radium. It was isolated from radium chloride, and she determined that its atomic weight was 225.93. Before Marie began her almost obsessive work in isolating radium, it would have been difficult to isolate the individual contributions of Marie and Pierre. However, in 1899, the division of labor became more apparent, Marie assuming the role of a chemist and Pierre that of a physicist. *See also* POLONIUM; RADIOACTIVITY.

RADIUM, HEALTH PROBLEMS FROM. From their earlier work on **radium**, both **Marie** and **Pierre Curie** were cognizant of some of the ills that radium might cause. Marie occasionally had suppurating sores on her fingers, and workers in her laboratories complained of fatigue when working in an atmosphere of **radon**. They also observed that **radioactive** elements apparently had different effects on different people. For example, Pierre Curie had excruciating pain in his legs whereas Marie, working under the same conditions, had only a few symptoms of lethargy. Later she began to have pains in her arms. Pierre Curie demonstrated that cells of animal tissues were destroyed after exposure to large doses of radium. Workers in Marie's laboratory in the 1920s became aware of some of the problems of working with radioactive substances, but they took only minor precautions such as changing laboratory coats frequently. However, reports from London in the 1920s disturbed the laboratory workers. **Ellen Gleditsch** expressed her concerns to Marie Curie and explained that Norway had set up a committee to investigate and that she was a member. In the early 1920s, many radium workers in Europe and North America complained of alarming symptoms. In 1924, a New York dentist, Theodor Blum, had several patients with cancerous jaws. They had been misdiagnosed as having syphilitic osteomyelitis. Blum found that these patients had all worked in watch-dial painting. When painting luminous figures on the watch dials, they would lick the camel-hair brush, point it with their tongues, and then apply the paint. When reports of this situation came to Marie Curie's attention, she recalled many of her own symptoms such as low blood pressure and anemia. Still she faced a dilemma. It seemed that radium, her "baby" had the power to both cure and kill. After World War I ended, she was lauded for her work in using radium to cure cancer. This realization led funding agencies to provide money for cancer research in her laboratory. She was hesitant to publicly recognize that radium studies acted as a two-edged sword, but after the watch-dial situation and increased evidence of sickness on the part of laboratory workers she apparently recognized the dilemma. The radiation that killed the cancer cells also attacked normal cells. *See also* HEALTH PROBLEMS OF M. CURIE; HEALTH PROBLEMS OF P. CURIE; MELONEY, MARIE MATTINGLY.

RADIUM, INTERNATIONAL STANDARD FOR. It was important for any research worker involved with making measurements using **radium** to know the purity of the sample being used. Physicians who used radium to treat patients in hospitals needed to know the exact amount to be used. Consequently, it was vital to establish an international standard for a precisely known amount of radium. Then secondary standards could be prepared for individual countries. However, national rivalries often intrude on international standards of any kind, and radium was no exception. Although national rivalries did play a part, most finally agreed that **Marie Curie**'s eminence gave her the right to establish the standard. The standard was established in 1911, and a small tube was deposited at the **International Bureau of Weights and Measures near Paris**.

RADIUM INSTITUTE. When Marie was recovering from her illness following the Langevin scandal, she considered an offer to establish a **radium** laboratory in **Poland**, far from the vitriolic French press. Although she visited **Warsaw** in 1913 for the inauguration of the building, she decided to direct the laboratory from Paris rather than move to Warsaw. **Marie** and **Pierre Curie** had dreamed of a Radium Institute in Paris, and as she realized that the Paris laboratory she and Pierre had discussed had been approved in 1909 she felt that she must stay in order to honor Pierre. The process of approval had been complicated. After Pierre's death, public authorities had proposed a national subscription to build a Curie Institute. Marie refused because she did not want Pierre's death turned into money; consequently, nothing was done. However, in 1909, the director of the **Pasteur Institute** wanted to build a laboratory for Marie Curie that would involve her leaving the **Sorbonne**. The university refused to let that happen, and a compromise was agreed upon. Together the university and the Pasteur Institute would share the expense. The institute would have two parts: a radiological research laboratory directed by Curie, and a laboratory for biological research on the practical applications of radiological research to medicine, headed by the physician **Claudius Regaud**. The two institutions would be constructed next door to each other and would be known as the Institut du Radium. The Radium Institute begun in 1912 was completed during World War I. Although Marie had never shown any interest in decorating her own home, when it came to the laboratory she was proactive, interested in every detail. By the 1920s, the two original laboratories had expanded. Curie's laboratory had doubled, and the number of researchers had increased from just a few to between 30 and 40, including a number of **women**. *See also* CURIE–LANGEVIN SCANDAL.

RADON GAS. Radon gas is produced by the natural disintegration of radioactive heavy metals such as **uranium** and **thorium**. As the atoms or the heavy metals disintegrate, they change into lighter radioactive heavy metals until they finally end up with stable nonradioactive lead. As an example, uranium becomes thorium, thorium decays into **radium**, radium into radon, radon into **polonium**, and polonium eventually becomes lead. Bad publicity occurred when this radon gas was found in homes in the United States. As a carcinogen, it causes lung cancer. This emanation is what **Marie Curie**'s followers used to treat cancer.

RAMSAY, SIR WILLIAM (1852–1916). The Scottish chemist Sir William Ramsay discovered the noble gases (argon, helium, neon, krypton, and xenon) and received the Nobel Prize in Chemistry in 1904. When **Lord Kelvin** rejected the idea of the transformation of one element into another as alchemy, Ramsay was one of three distinguished scientists who claimed transformation in other elements. He reported that when **radon** was combined with copper, the copper began to disintegrate in much the same way that radioactive elements did. He reported that when he combined radon and copper it produced lithium, an element that is in the same series as copper but with a lower atomic weight. **Ernest Rutherford** was suspicious but could not repeat Ramsay's experiment because he did not have the **radium**. When **Marie Curie** tried to reproduce his results, she found that the glass tubes in which Ramsay had combined radium and copper sulfates introduced lithium into his final product. When they used lithium-free containers, the amount of lithium in the product was infinitesimal. Ramsay was not a fan of Curie, nor she of him. He said of Curie and **Hertha Ayrton** that **women** scientists do their best work when collaborating with a male. When he began to work in her area of **radioactivity**, she claimed that in his work on the atomic weight of radium he arrived at exactly the same results as she had found earlier, but he claimed that his work was the first good work on the subject. Curie was furious and claimed that Ramsey had made incorrect comments about her experiments on atomic weights.

RAMSTEDT, EVA JULIA AUGUSTA (1879–1974). The Swedish physicist Eva Ramstedt was educated at the University of Uppsala (B.A., 1904; licentiate, 1908; Ph.D. 1910). She was one of the women employed in **Marie Curie**'s laboratory, studying radium decay products during 1910–1911. She found that the behavior of the solid products depended on the surface upon which they were collected, and that the solubility of radium emanation (**radon**) varied with the solvent used and the temperature. She returned to Sweden and worked at the Nobel Institute under Svante Arrhenius. She published the results of her studies. She was very involved in the education of **women**. Two of her publications were about Marie Curie. *See also* WOMEN SCIENTISTS IN MARIE CURIE'S LABORATORY.

RAYLEIGH, LORD (STRUTT, JOHN WILLIAM) (1842–1919). John William Strutt, Third Baron Rayleigh, was a British physicist who contributed to both theoretical and experimental physics. He won the Nobel Prize in Physics (1904) and was one of the elite of British science that **Marie** and **Pierre Curie** met when they were in England in 1903.

RECHERCHES SUR LES SUBSTANCES RADIOACTIVES. French title of **Marie Curie**'s **doctoral dissertation**. *See also* RESEARCHES ON RADIOACTIVE SUBSTANCES.

REGAUD, CLAUDIUS (1870–1940). Claudius Regaud was a French physician and biologist who was a pioneer in radiotherapy at the Curie Institute. In his research, he found that X-ray treatment could cause sterility and that it could affect other rapidly growing cells such as cancer. *See also* PASTEUR INSTITUTE; RADIUM INSTITUTE.

RESEARCH OF P. CURIE. Pierre Curie had already made a name for himself in his research on crystals when he interrupted his own investigations to work on **radioactivity** with his wife. He never really returned to his earlier research. The couple used two of Pierre's inventions, the **electrometer** and the **piezoelectric quartz balance**, to determine whether other substances besides **uranium** caused the air to conduct electricity. Marie and Pierre developed a technique to test various substances using the electrometer.

RESEARCHES ON RADIOACTIVE SUBSTANCES. The title of the translation of the publication of **Marie Curie**'s dissertation. After Marie Curie produced one-tenth of a gram of almost pure **radium** prepared from radium chloride and determined its atomic weight (225.93), she wrote her **doctoral dissertation**, *Researches on Radioactive Substances*, which she defended on 25 June 1903. She was awarded the degree of Doctoral of Physical Sciences at the **Sorbonne** (University of Paris) with the added accolade of *tres honorable*. This dissertation was published the following year.

ROGOWSKA, ANNA MARIA. Maria Sklodowska spent the summer after she graduated from **Gymnasium Number Three** visiting her relatives in the country. Anna Maria Rogowska was the wife of Maria's father's brother Zdzislaw. *See also* ROGOWSKI, ZDZISLAW.

ROGOWSKI, ZDZISLAW. Zdzislaw Rogowski was Maria Sklodowska's uncle, the brother of her father and the husband of **Anna Maria Rogowska**. Maria spent the latter part of her vacation after graduating from **Gymnasium Number Three** at their country home. The Rogowskis had three daughters, and the family often had parties that Maria enjoyed greatly.

RÖNTGEN, WILHELM (1845–1923). A Nobel Prize winner in 1901, Röntgen is best known for his discovery of X-rays. As **Marie Curie** was searching for a topic for her **doctoral dissertation**, she was perusing the scientific literature and became aware of Wilhelm Röntgen's discovery of these rays. Röntgen realized that a conductive material (electrode) used to make electrical contact with part of a circuit could be charged positively (**anode**) or negatively (**cathode**). The cathode is the source of **electrons**. Röntgen investigated the properties of cathode rays (electrons) emitted by a high-vacuum discharge tube. He noted that when activated by a high-voltage current, the electrons would race from the cathode to the anode. Although often the electrons (cathode rays) were invisible, sometimes they appeared as blue streaks. If the streaks (cathode rays or electrons) touched the glass wall of the tube, they emitted a green or blue luminescence. Röntgen was fascinated by the work of **Philipp Lenard** and tried to repeat his experiment. He wrapped the cathode tube in black cardboard in order to exclude light in a darkened room so that he could see a faint light. He activated the tube and observed a flash of light but, contrary to his expectations, found that the light did not come from the tube. He observed that a sheet of barium platinocyanide glowed (phosphoresced) even when the tube, blocked off by the black cardboard, could not possibly have reached the barium platinocyanide. When he turned the coated part of the paper screen away from the discharge tube, it still phosphoresced. He even carried the coated paper into another room and it still glowed. He found to his surprise that as long as the discharge tube was still in operation, the paper would continue to glow. He postulated that the rays that passed unchanged through cardboard and thin plates of metal were not deflected by electric or magnetic fields, as were cathode rays. He discovered a new type of radiation, and named it X-rays. These rays actually came from the glass walls of the tube when struck by cathode rays and possessed the ability to penetrate solid substances. This discovery excited both scientists and laypersons and led to a search of other hitherto unknown rays. Marie Curie was intrigued by this new ray that Röntgen and later **Henri Becquerel** described and was inspired to continue the work for her dissertation. See also PHOSPHORESCENCE.

ROUX, PIERRE PAUL ÉMILE (1853–1933). Émile Roux was a French physician, bacteriologist, and immunologist and was a close collaborator with Louis Pasteur (1822–1895). Beginning in 1883, Roux became involved with the creation of the **Pasteur Institute**, where he divided his time between administrative duties and biomedical research. He was instrumental in the founding of the **Radium Institute** in collaboration with the **Sorbonne**.

ROYAL GREENWICH OBSERVATORY. This observatory, as well as the **Harvard College Observatory**, provided paid positions for **women** scientists. Although the pay for women was low and greatly benefited the observatories, it still provided women with scientific paid work outside the home.

ROYAL INSTITUTION, LONDON. The Royal Institution was founded in 1799 as the result of a proposal by American-born scientist Benjamin Thompson, Count Rumford. From its inception to the present day, it has supported public engagement with science through a program of public lectures. One of the most famous of these is the Royal Institution Christmas Lectures, founded by Michael Faraday in 1825. In 1903, **Pierre Curie**, accompanied by Marie, made a trip to London for Pierre to give an invited lecture at the Royal Institution. While Pierre was lecturing, Marie, who had done much of the work he described, sat in the audience. The impression given was that Pierre was the important scientist and Marie his assistant. Pierre, however, did not intend for the audience to reach that conclusion and carefully acknowledged Marie's essential work in the collaboration. On that trip, Marie met an important fellow woman scientist, **Hertha Ayrton**, and her husband.

ROYAL SOCIETY OF LONDON. The prestigious Royal Society was founded in the 17th century (November 1660) and is the oldest national scientific institution in the world. Influenced by the *New Atlantis* of Francis Bacon, the precursors of the society first met in various locations, including Gresham College, London. The society was formalized on 28 November 1660, when a committee of 12 announced the formation of a "College for the Promoting of Physico-Mathematical Experimental Learning," which would meet weekly to discuss ideas in science and give experimental demonstrations for its members. A royal charter was signed on 15 July 1662, with Lord Brouncker as the first president. A second royal charter, signed on 23 April 1663, noted the king as founder and gave it the name of the Royal Society of London for the Improvement of Natural Knowledge. Robert Hooke was appointed curator of experiments. The fellows of the society included notables such as Sir Isaac Newton. In June 1903, **Marie Curie** learned that the Royal Society had awarded both she and **Pierre Curie** the **Davy Medal**, given out annually for the most important discovery in chemistry. Marie was still recovering from a miscarriage and did not accompany Pierre to receive the medal. Pierre stayed with Sir **William Huggins** and Lady **Margaret Huggins** while he was in London accepting the medal. In November of that year, Marie and Pierre Curie were informed in confidence that they with **Henri Becquerel** had won the **Nobel Prize in Physics**.

RUSSIA. Russia dominated the **Poland** of **Marie Curie**. Poland's government had long been chaotic and corrupt and was ripe for a takeover by its stronger and more efficient neighbors, Austria, Russia, and Prussia. Three partitions of the Polish–Lithuanian Commonwealth resulted in the elimination of sovereign Poland and Lithuania for 123 years. The partitions were conducted by Habsburg Austria, the Kingdom of Prussia, and the Russian Empire, which divided Polish lands between them. The rise of Napoleon seemed to give the Poles an opportunity to free themselves; however, this hope was squelched when Napoleon was disastrously defeated by Russia (1811–1812). Napoleon's Duchy of **Warsaw** was replaced by a Kingdom of Poland connected to Russia by a union with the tsar of Russia. The tsar also became king of Poland, which had its own constitution, parliament, army, and treasury. The Prussians governed the remaining territories. There were constant tensions between Russia and the constitutional regime in Poland. There was a rebellion in 1830, but the Poles could not prevail against the superior Russian resources, and many of the concessions the Poles had gained previously from Russia were taken away. Another failed uprising occurred in January 1863 and lasted through 1865. After this rebellion, the Poles gave up the idea of defeating Russia by military means and resorted to civil disobedience instead. Russian authorities and the Poles realized that they needed the support of the peasants in order to succeed. In 1864, the Russian tsar issued a decree enfranchising the peasants. However, this new freedom did not have the effect that the Russians expected, for the peasants gradually became members of the National Polish Community, which was the goal of the rebellious Poles. A Poland dominated by Russia was the Poland where Marie Curie grew up. Russian influence was everywhere in the schools.

RUTHERFORD, ERNEST, FIRST BARON RUTHERFORD OF NELSON (1871–1937). New Zealand–born British physicist Ernest Rutherford, who had studied with **J. J. Thomson** at Cambridge, is known for discovering the concept of **radioactive** half-life, the radioactive element **radon**, and difference between alpha and beta radiations. He received the Nobel Prize in Chemistry (1908) for his work performed at McGill University in Canada. He left McGill for the Victoria University of Manchester (now University of Manchester) in 1907. Here he theorized the model of the atom that is basically accepted today through interpreting the gold foil experiment of **Hans Geiger** and **Ernest Marsden**. Rutherford suggested that they shoot alpha (positively charged) particles at a thin sheet of gold, assuming that the particles would go straight through the foil with little deflection. The contemporary model of the atom (known as the plum pudding model) assumed that the negative electrons (the plums or raisins) would be spread evenly throughout the positive matrix (the pudding). To their surprise, although 98 percent of the particles went straight through the foil, about 2 percent bounced off the gold foil. Since **alpha particles** have about 8,000 times the mass of an **electron**, it would take a very strong force to deflect the particles. Rutherford interpreted these results to mean that most of the atom's mass was concentrated into a compact positive nucleus with electrons occupying most of the atom's space. In 1917, he discovered the first artificially induced nuclear reaction by bombarding nitrogen nuclei with alpha particles that resulted in the emission of a subatomic particle, which in 1920 he named the proton. He became director of the Cavendish Laboratory at the University of Cambridge in 1919 and was knighted in 1914. During his career, he received numerous honors. **Marie Curie**'s relationship with Rutherford was often strained. In his correspondence with **Bertram Boltwood**, he was less than flattering about her contributions to science. However, when Curie was in the middle of the **Paul Langevin** affair, Rutherford was supportive of both his longtime friend Langevin and Marie Curie. In an obituary in the British journal *Nature*, he was more positive toward Curie than he had been during her lifetime.

S

SAGNAC, GEORGES (1869–1928). Sagnac was a French physicist who was one of the first Frenchmen to study X-rays after **Wilhelm Röntgen**'s discovery. He belonged to a group of friends and scientists that included **Pierre** and **Marie Curie**, **Paul Langevin**, Jean Borel, and Émile Borel. Pierre even produced a paper with Sagnac on X-rays and **radioactivity**. Their friendship also had its personal side. After the discovery of **radium** in 1898, both Curies were fatigued. Sagnac suggested that they must eat correctly at regular hours, must not talk physics when they were eating, and warned them that they were not getting enough rest. Although these suggestions were clearly good ones, most of the health issues that they had at that time we now know are attributable to exposure to radiation.

SANCELLEMOZ SANATORIUM. In 1934, **Marie Curie**'s health declined precipitously. Parisian doctors observed old tubercular lesions on X-rays and recommended that she be taken to a sanatorium, Sancellemoz, in the Savoy mountains. The trip to the sanatorium by train was a difficult one. **Eve Curie** accompanied her mother, and she was placed in the best room in the sanatorium. The new X-rays showed no sign of tuberculosis, and the Swiss doctor who examined her blood tests diagnosed her as having an extreme form of aplastic pernicious anemia. On 4 July 1934, Marie Curie died at the sanatorium. *See also* HEALTH PROBLEMS OF M. CURIE.

SCHMIDT, GERHARD CARL (1865–1949). The chemist Schmidt was born in London to German parents. His discovery that **thorium** was **radioactive** occurred 19 days before Marie announced her discovery to the **Académie des Sciences**.

SÈVRES. *See* ÉCOLE NORMALE SUPÉRIEURE DE JEUNES FILLES.

SÈVRES, TEACHING POSITION AT. *See* ÉCOLE NORMALE SUPÉRIEURE DE JEUNES FILLES.

SIKORSKA, JADWIGA (MADAME) (1846–1927). Educator Jadwiga Sikorska took a one-year pedagogical course and obtained a certificate that entitled her to teach Polish in private secondary and government schools. After teaching in several educational institutions, she obtained permission from the school district curator to open a private female school in **Warsaw**. This private school was attended by both Maria and **Helena Sklodowska**. Although Sikorska was required to teach in Russian only, through an elaborate cover-up she actually taught the Polish language, geography, and history. *See also* STUDENT, M. CURIE AS A PREUNIVERSITY.

SKLODOWSKA, BRONISLAWA (BRONIA). *See* DLUSKA, BRONISLAWA.

SKLODOWSKA, BRONISLAWA. **Marie Curie**'s mother, Bronislawa Boguska, married her father, **Wladyslaw Sklodowski**, in 1860. Even though the Boguski family belonged to the landowning nobility, its members had financial problems. They were, however, able to give their daughter, Bronislawa, a good education in a private school in **Warsaw**. Although she lacked money, she was well educated, accomplished, beautiful, and musical. When she married Wladyslaw, she was a teacher in the same private school that she had attended and eventually became its director. The couple first lived in apartments adjacent to Bronislawa's classrooms for seven years and had five children during this time, Zofia, Józef, Bronislawa, Helena, and Maria. Later, they moved into quarters supplied by Wladyslav's teaching position. Bronislava experienced the first symptoms of tuberculosis when Maria was just a baby. Her oldest daughter, Zofia (Zosia), took over much of the responsibility for her mother's care, including accompanying her to a spa, where she served as her mother's nurse. After two years at the spa, mother and daughter returned to their home. Political events caused Wladyslaw to lose his positions, and the family lost its living place. In order for the family to survive, Wladyslav turned their new home into a boarding school. A typhus epidemic swept through the school and afflicted Zosia and Bronia. Fourteen-year-old Zosia died of the disease. Bonislava, the mother, never recovered from the death of Zosia but did not die until two years later, on 9 May 1878, from tuberculosis.

SKLODOWSKA, HELENA (HELA) (1866–1961). Helena was the sister closest in age to Maria Sklodowska. Helena and Maria were in the same class at Madame **Jadwiga Sikorska**'s private school, although Helena was a year older. Known as Hela by her family and friends, she became a teacher and educational activist in **Poland**. She married Stanislaw Szalay (1867–1920), who was a **Warsaw** photographer. The couple had one daughter, Hanna. She was buried in the Powazki Cemetery in Warsaw. She published her memories of her sister, **Marie Curie**, in 1958, *Ze Wspomnienia o Maria Sklodowskiej-Curie*.

SKLODOWSKA, MARIA (MANYA) (1867–1934). *See* CURIE, MARIE.

SKLODOWSKA, ZOFIA (ZOSIA) (1861–1876). Known to her family as Zosia, Zofia Sklodowska was the oldest of the five children of Bronislawa (Boguska) and **Wladyslaw Sklodowski**. When their mother, **Bronislawa Sklodowska**, became very ill with tuberculosis, Zosia as the oldest daughter was charged with her mother's care. She accompanied her mother to the spa where Bronislawa went for treatment and served as her nurse, maid, laundry woman, and entertainer. Her father assumed that since daughter Zosia, too, was considered "delicate," the treatments at the spa such as sulfur baths and long naps would benefit her. He did not seem to consider the danger of exposing Zosia to tuberculosis. After two years at the spa and as Bronislawa was not improving, the two moved back home. Zosia was an excellent student and like several of her siblings graduated first in her class. The situation at home was dire because with the increased Russianization in **Warsaw** father Wladyslaw lost his job and home. In order for the family to survive, Wladyslaw turned their new home into a boarding school for boys. The overcrowded facilities may have contributed to the typhus that Zosia and Bronia contracted. Although Bronia recovered, 14-year-old Zosia, the favorite of her mother, succumbed to the disease. Two more years passed before Bronislawa finally died of tuberculosis.

SKLODOWSKI, JÓZEF (1863–1937). Józef Sklodowski was Maria Sklodowska Curie's brother. Józef received a gold medal and studied at the University of **Warsaw**. His sisters were envious because they, as girls, were unable to attend this university. He became a physician in **Poland**. When **Marie Curie** was excoriated during the **Curie–Langevin affair**, Józef and Bronia left Poland for Paris to support their sister and unsuccessfully tried to convince her to return to Poland. He updated the family history written by Marie's father, **Wladyslaw Sklodowski**, with information on his generation. An extensive correspondence is available between Józef and Marie. Józef was married twice. His first wife was Jadwiga, with whom he had two daughters, Janina and Maria, and a son, Wladyslaw. His second wife was named Maria.

SKLODOWSKI, WLADYSLAW (1832–1902). Wladyslaw Sklodowski, who married Bronislawa Boguska in 1860, was the father of five children: Zofia, Józef, Bronislawa, Helena, and Maria. He graduated from high school in Siedlce with a gold medal, setting a precedent for his children. He studied science at the **University of St. Petersburg** in Russia and then returned to **Warsaw**, where he taught mathematics and physics. After his marriage, he and **Bronislawa Sklodowska** moved into apartments adjacent to her classrooms. For several years, he managed to convince the Russian officials that he was not disloyal to the regime, although he was always suspect. When the policy called Russianization developed, in which Polish officials were replaced by Russian immigrants, Sklodowski lost his teaching job, his job as underinspector, and the family home (because it was furnished by the school). In order to survive, he turned his new home into a boarding school for boys. Maria developed a love of science from her father, who encouraged her in her favorite studies, mathematics and physics. After her marriage to **Pierre Curie**, the couple visited her father in Warsaw several times. The last visit was in 1899, when the entire family reunited in Zakopane in the Carpathian Mountains. Shortly after that visit, he was hit by a truck and had a debilitating fracture. Although he recovered somewhat from the accident, shortly thereafter he had a gallbladder attack and had surgery to remove large gallstones. He died before the train that Maria was on arrived at the Warsaw station.

SOCIETY FOR THE ENCOURAGEMENT OF NATIONAL INDUSTRY. As **Marie Curie** was completing her mathematics degree, she was hired by the Society for the Encouragement of National Industry, formed to promote French science. One of her jobs was to study the magnetic properties of various steels. However, she did not have access to a laboratory in order to do the work. As she was searching for laboratory space, Józef Kowalski, a Polish physicist, suggested that she meet with **Pierre Curie**, who was working on magnetism and might have space available. When they met, they found many common interests, but neither expected a romance to occur. *See also* EARLY RESEARCH OF M. CURIE.

SODDY, FREDERICK (1877–1956). Soddy was a radiochemist who, along with **Ernest Rutherford**, postulated that **radioactivity** is due to the transmutation of elements. He worked with Rutherford at McGill University in Montreal, Quebec. They realized that the strange behavior of radioactive elements resulted from their decay into other elements.

SOLVAY, ERNEST (1838–1922). Belgian chemist and industrialist Ernest Solvay had amassed considerable wealth from his patents. This money allowed him to bankroll several philanthropic organizations, one of which was the Solvay Institutes. He is best known for developing a commercially viable process for producing soda ash (sodium carbonate), used in the manufacture of products such as glass and soap. Although the process had been long understood, Ernest and his brother Albert solved the problems of large-scale production. He and Albert founded their own company in 1863 and had a factory built. They soon expanded to other countries and amassed considerable wealth, funding various international institutes of scientific research in chemistry, physics, and sociology, including the **Solvay Conferences**.

SOLVAY CONFERENCES. The first Solvay Conference was held in Brussels from 30 October to 3 November 1911 on the subject *La théorie du rayonnement et les quanta* (Radiation and the Quanta). This conference considered the difficulty of having two different approaches to physics: classical physics and quantum physics. Many luminaries were in attendance, including **Albert Einstein** (the youngest physicist present), **Marie Curie**, Max Planck, **Ernest Rutherford**, and **Paul Langevin**. While Marie Curie was attending this conference, the story of her alleged affair with Paul

Langevin broke in the newspaper *Le Journal*. Marie Curie attended the Solvay Conferences of 1911, 1912, 1913, 1921, and 1933. *See also* CURIE–LANGEVIN AFFAIR; SOLVAY, ERNEST.

SORBONNE. During the 12th century, universities arose from cathedral schools in several different locations, including Bologna, Oxford, and Paris. The University of Paris, later known by its nickname, the Sorbonne, after its theological College of Sorbonne founded by Robert de Sorbon, was established about 1150 and was associated with the cathedral school of Notre Dame de Paris. It was officially chartered in 1200 by King Philip II of France and recognized in 1215 by Pope Innocent III. In 1793, during the French Revolution, the university was closed and the college endowments and buildings sold. Napoleon replaced it with a new University of France in 1806, with four independent faculties. The defeat of France during the Franco-Prussian War in 1870 was often partially blamed on the superior Prussian education system and led to the reform of the university, and **Marie Curie** was a beneficiary of these reforms. After a series of student revolts beginning in 1966 and ending in 1970, the old university was closed and replaced by 13 successor universities, one of which was the Pierre and Marie Curie University (Paris 6). *See also* SORBONNE, M. CURIE AS A PROFESSOR AT THE; SORBONNE, M. CURIE AS A STUDENT AT THE.

SORBONNE, M. CURIE AS A PROFESSOR AT THE. After **Pierre Curie**'s death, his position at the **Sorbonne** was vacant, and the question arose as to who would take over his teaching. The university agreed to give Pierre's widow, **Marie Curie**, a pension, but she refused it, stating that she was too young to accept a pension and that she could support herself and her children. Marie's family and friends took the initiative and informed the dean that she was the only French physicist competent to succeed Pierre. No **woman** had ever held such a position, but the council of the Faculty of Science decided unanimously to offer her an assistant professorship. She also received the chair that had been especially created for Pierre and which he had only occupied for 18 months. Several hundred people gathered for her first lecture, curious to see how the first woman professor would perform. Some attendees hoped for drama and others willed her to fail because they were convinced that a woman should not hold such an exalted position. However, Curie began at the exact sentence where Pierre had left off. "When one considers the progress that had been made in physics in the past ten years, one is surprised at the advance that has taken place in our ideas concerning electricity and matter."

SORBONNE, M. CURIE AS A STUDENT AT THE. In March 1889, **Marie Curie** registered at the **Sorbonne**, one of the oldest universities in the world. Although the university had been a bulwark of traditional church doctrine, by the time that Curie had arrived improvements had been made in the science curriculum —theology was banished, the humanities de-emphasized, and the sciences stressed. Her teachers included **Gabriel Lippmann**, who brought the German laboratory perspective to Paris; **Joseph Boussinesq**, who taught her classical physics; and **Henri Poincaré**, who made important contributions to mathematical theory and celestial mechanics. She was one of a very few **women** students, and the ones who were there were foreigners like herself. She, as a foreign student, had much more freedom than a French woman. Shy Marie had a difficult time with her relationship with other students, although she finally admitted that some of the students wanted to be friendly. She made friends with Polish students, and they often met socially. However, after her first year, she devoted all of her vacation time to study. It was especially difficult for her because not only did she have to master the subject matter, but she also had to take the examination in a foreign language. Her study paid off, for in 1893 she not only passed the *licence* examination in physics (a step beyond a bachelor of science degree) but received a first. To her father's dismay (he had assumed she would return home to **Poland** after her examination), she realized how important a mathematics background was to physics and chemistry and decided to work on an additional degree in mathematics. At the end of the next year (July 1894), she got a *licence* in mathematics and received a second degree. She defended her thesis in 1903 and received a Ph.D. degree. After **Pierre Curie**'s death, Marie became the first woman professor at the Sorbonne on 5 November 1906. *See also* SORBONNE, M. CURIE AS A PROFESSOR AT THE.

STOCKHOLM. The Nobel Prizes in Chemistry, Literature, Peace, Physics, and Physiology or Medicine were first awarded in 1901. With the exception of the peace prize, which is awarded in Oslo, Norway, the other prizes are awarded in Stockholm, Sweden. **Marie** and **Pierre Curie** did not go to Stockholm to accept their joint Nobel Prize (physics) with **Henri Becquerel** in 1903. Ill health may have been the reason, but they blamed their teaching schedules for their absence. Becquerel, however, did attend. In June 1905, the Curies traveled to Sweden, and Pierre gave the lecture required of Nobel laureates. Marie Curie, accompanied by sister Bronia and daughter Irène, attended the 1911 Nobel Prize ceremony for her Nobel Prize in Chemistry and gave the lecture. *See also* NOBEL PRIZE FOR CHEMISTRY, 1911, M. CURIE; NOBEL PRIZE FOR PHYSICS, 1903, M. AND P. CURIE AND H. BECQUEREL.

STOKES, GEORGE (1819–1903). George Stokes was an Anglo-Irish physicist and mathematician. He received the Copley Medal of the **Royal Society** and served as president of that society from 1885 to 1890. He spent his entire career at Cambridge University (Lucasian Professor of Mathematics from 1849 to 1903). His connection to the Curies was through **Silvanus P. Thomson**, who independently duplicated **Henri Becquerel**'s results. Thompson wrote to Stokes, who was president of the Royal Society at that time, to report his results. Stokes was enthusiastic and urged Thompson to publish immediately. Unfortunately for Thompson, Becquerel published first and received the credit.

STONEY, GEORGE JOHNSTONE (1826–1911). G. J. Stoney was an Irish physicist who is best known for introducing the term *electron* (at first he called it "electrine") for "the fundamental unity of electricity." The atom (from the Greek *atomos*, meaning "uncut") had long been considered the fundamental unit of matter. **J. J. Thomson** had made the first "cut" in the "uncuttable" atom after he postulated that negative **cathode** rays were streams of particles much smaller than atoms. He concluded that these light particles were universal constituents of matter and called them "corpuscles." These "corpuscles" were later given the name electrons first used by Stoney.

STRUTT, JOHN WILLIAM (LORD RAYLEIGH) (1842–1919). Physicist Lord Rayleigh, the winner of a Nobel Prize for Physics in 1904, was one of the elite members of British society that **Marie** and **Pierre Curie** met when they visited England in 1903.

STUDENT, M. CURIE AS A PREUNIVERSITY. Marie's formal education began in the Freta Street school in the building where her mother, now desperately ill with tuberculosis, had been headmistress. The next year her parents decided to enroll her and her sister Helena in a school closer to home, the private school of Madame **Jadwiga Sikorska**. During this time, the schools in **Poland** were required to teach only in Russian and were subject to punishment if they were caught using Polish. Madame Sikorska was able to conceal what she was really doing, teaching the Polish language, geography, and history. Although some of the inspectors were benign, others were quite menacing. The students and teachers were engaged in an elaborate subterfuge to cover up what they were really teaching and learning. Although Maria was the youngest student in the class, she was also the brightest. Her well-educated father inspired her love for science and for literature, especially poetry. However, her favorite subjects were mathematics and physics. After Maria's mother's death, Madame Sikorska realizing that Maria was emotionally distraught suggested to her father that she stay out of school for a year. He rejected her advice and enrolled her in a government-run advanced high school (**Gymnasium Number**

Three) in downtown **Warsaw** in order to expose her to a more rigorous education. It is difficult to understand why he did this at this time. He knew that government institutions were staffed by unqualified or uninterested teachers who enforced the use of Russian in the classroom. Although Sklodowski had advocated delaying **gymnasium** for his children as long as possible, for some reason he enrolled his youngest daughter into the girls' Gymnasium Number Three. It proved to be better than the corresponding boys' gymnasium because before Russian occupation it had been a German-speaking school, with German respect for learning. It had better teachers than the boys' school, including a good physics teacher. Still, Maria's memories of the school were grim.

SZALAY, HELENA (HELA). *See* SKLODOWSKA, HELENA.

T

LE TEMPS. After *Le Journal*, a Paris daily newspaper published from 1892 to 1944, had published its scathing report of letters between **Marie Curie** and **Paul Langevin**, Marie wrote a letter to the newspaper *Le Temps*, which had been tepidly supporting Curie, excoriating the press for meddling in her private life. She threatened to demand monetary damages. The *Journal* apologized and retracted the story. After Marie's letter and the reporters' denial were widely publicized, it seemed like the accusations would disappear.

TÉRY, GUSTAVE (1871–1928). Téry was the founder of the right-wing newspaper *l'Oeuvre*. *See also* CURIE–LANGEVIN AFFAIR.

THIRD REPUBLIC (FRANCE). The French Third Republic was the government adopted in France after the Second French Empire collapsed during the **Franco-Prussian War** in 1870 when Napoleon III was defeated. The republic lasted until 10 July 1940, when France was defeated by Nazi Germany in World War II and replaced by the Vichy government. The defeat of France in the Franco-Prussian War caused the French to blame their loss on the fact that their universities were no longer providing the best education in the sciences. This fear led them to reexamine their own education institutions and attempt to compete with the German institutions. **Marie Curie** reaped some of the results of this reform in her years at the **Sorbonne**. *See also* SORBONNE, M. CURIE AS A STUDENT AT THE.

THOMPSON, SILVANUS P. (1851–1916). Thompson found similar results to **Antoine-Henri Becquerel**, but Becquerel published first and got the credit. Thompson put a small amount of **uranium** nitrate over an aluminum-covered photographic plate and observed the effect. He developed the plate after putting it on a window sill. He was surprised to find that it had darkened at the place where the uranium salt had been. He was surprised that uranium could affect the plate through the aluminum shield. He called this phenomenon "hyperphosphorescence." *See also* STOKES, GEORGE.

THOMSON, JOSEPH JOHN (1856–1940). Nobel laureate physicist J. J. Thomson is known for "cutting" the "uncuttable" atom. He demonstrated that **cathode** rays were composed of previously unknown charged particles, which he called "corpuscles," and were determined to be over 1,000 times smaller than the atom. The term *electron*, used previously by **G. J. Stoney** (1826–1911), was accepted eventually for this negative particle. Thomson imagined that the atom was made up of these particles orbiting in a sea of positive charge, the so-called plum pudding model of the atom. His student **Ernest Rutherford** proved that this model was incorrect. Thomson was appointed Cavendish Professor of Physics in 1884 and received many awards, including the Nobel Prize for Physics (1906) and the Copley Medal of the **Royal Society** (1914). He was elected a fellow of the Royal Society in 1884 and served as its president from 1915 to 1920. After **Marie Curie**'s death, he characterized her as one of the greatest physicists of modern times.

THOMSON, WILLIAM (LORD KELVIN) (1824–1907). See KELVIN, LORD.

THORIUM. The naturally occurring slightly **radioactive** metal thorium was discovered in 1829 by the Swedish chemist **Jöns Jacob Berzelius**, who named it after the Norse god of thunder, Thor. When **Marie Curie** was working with **pitchblende**, known as a source of **uranium**, she found that it produced a current much stronger than that of uranium itself. **Henri Becquerel** had assumed that the "rays" were only produced by uranium. She tested different minerals with known elemental contents. Curie postulated that substances other than uranium produced these "rays." She found that a mineral, aeschynite, that contains thorium but no uranium also was more active than uranium. These observations led her to conclude that the "rays" were produced by substances other than uranium. **Pierre Curie** became interested in Marie's results, and the two of them postulated that the strength of the current produced by the pitchblende was caused by an unknown element that produced much more energy than the uranium itself.

TREATISE ON RADIOACTIVITY. This almost 1,000-page, two-volume work by **Marie Curie** was published in 1910 and was a comprehensive presentation of the progress in **radioactivity** since she began her observations in 1897. **Ernest Rutherford** reviewed it for *Nature* and gave it a favorable review, but in a letter to **Bertram Boltwood** he was patronizing.

TUPALSKA, ANTONINA. Tupalska was a math and history teacher at **Madame Sikorska**'s private school where Maria and **Helena Sklodowska** attended. Maria's mother was very ill with tuberculosis, and the children's father arranged for Antonina Tupalska to board with them and walk the children to school. A strict disciplinarian, "Tupcia," as the children called her, was not a physically attractive person and seemed to the children to try to usurp the position of their ill mother. Nevertheless, she was kindly and loved and was proud of the children. *See also* STUDENT, M. CURIE AS A PREUNIVERSITY.

TYPHUS. The Sklodowski family was forced to move after father Wladyslaw lost his teaching position and his position as underinspector as the Russian policy against the Poles became more severe. The family lost their living quarters and income. **Wladyslaw Sklodowski** turned their new dwelling into a boy's boarding school. The conditions at the crowded boarding school may have provided the source of the typhus that afflicted the family. Both Zofia (Zosia) and Bronia contracted typhus in 1874. Although Bronia recovered, 14-year-old Zosia, their mother's favorite, died. Although their mother lived two additional years, she never recovered from the death of her favorite daughter.

U

UNITED STATES, M. CURIE'S 1921 TRIP TO THE. Marie Meloney (Missy) was able to talk **Marie Curie**, whom she idolized, into visiting the United States in 1921 to take part in a campaign to raise funds from American women for a gift to Curie of a gram of **radium**. Meloney couched her campaign in terms of radium's ability to "cure" cancer. Although this campaign was launched during the period after the **Paul Langevin** scandal, Curie was still wary of the French newspapers, some of which were still hostile to her. She was concerned about her reception in the United States, but Meloney assured her that all would be well. Meloney arranged a whirlwind of activities, including conferences, honorary degree ceremonies, and award acceptances.

Curie and her daughters, Irène and Eve, would be met by representatives of 100,000 college **women**. The Association of College Alumnae and the Woman's University Club would host a reception, and each alumna would be asked to donate one dollar to the radium fund. Without Curie's concurrence, Meloney arranged for the Curies to stay for five weeks as a guest of the Marie Curie Radium Fund Committee. They were to visit eastern

Marie Curie in the 1920s

Marie Curie in the United States in 1921 with Dean George Pegram of Columbia University

cities, the Midwest, and even the Grand Canyon. Curie was to receive the gold medal of the National Institute of Social Sciences, recognizing her discovery of radium and its benefit to humanity. The three Curies and Meloney boarded the ship the *Olympic* on 7 May 1921, and the *New York Times* announced their arrival on 11 May. The Americans had not only collected their goal of $100,000; they had oversubscribed by about $50,000. Overwhelmed by the exuberant Americans, they went from event to event. On the day before the actual gift of the radium was to be presented by President **Warren Harding** on 20 May 1921, a problem arose. When Meloney read Curie the entire document aloud, she was displeased because a sentence Curie deemed vital had been left out of the document. It did not provide for the succession of the radium after Curie's death. She insisted that the document read that the radium was "for free and untrammeled use by her [Curie] in experimentation and in pursuit of knowledge" and that it would become the property of her laboratory after her death. She insisted that they have a lawyer process the deed of gift even though it was late at night when she realized the problem. The women who had worked so hard to collect the money were probably not happy when Curie stubbornly insisted on her own way. They still did not arrive at a satisfactory solution for the leftover money. Although her benefactors thought they should have some say in its disposition, Curie thought she alone should say how it should be used. Eventually, she triumphed. On 20 May 1921 at four o'clock, President Harding presented Curie with the gram of radium. After the presentation and subsequent "hospitality," Marie was exhausted and cut her journey short. Her trip to the West Coast was canceled, but she did see the Grand Canyon. The exuberance of Americans was almost more than Curie could manage. *See also* THE DELINEATOR.

UNITED STATES, M. CURIE'S 1929 TRIP TO THE. Marie Curie's physician sister Bronia had attempted to raise money for an **radium** institute in **Warsaw** with disappointing results. Even though the building was partially constructed, there was not enough money for the radium for cancer treatments. For the second time, **Marie Meloney** galvanized the American public to support a campaign to supply radium for **Poland**. The money was raised just before the stock market crash in the United States. Marie Curie arrived in the United States on 16 October 1929, just ahead of the crash. Although Curie would have preferred a degree of anonymity, the Americans would not let her hide. However, she made it clear that her physical limitations would limit what she could do. Nevertheless, her popularity made it difficult to turn down certain invitations. She and Meloney were guests of President **Herbert Hoover** and his wife and stayed in the White House for several days. The presentation of the radium gift of $50,000 was made on 30 October 1929. Ignoring the advice of her physicians and her own wishes, she attended dinners, receptions, and different forms of entertainment provided by her hosts. After celebrating her 62nd birthday on 7 November, she departed on the French liner *Ile de France* for Europe the next day. She arrived in France with little fanfare on 15 November 1929, tired but grateful for the hospitality she had received and for the gift of radium.

UNIVERSITY OF ST. PETERSBURG. Marie Curie's father, **Wladyslaw Sklodowski**, studied science at this university in Russia. It was established by a decree of Peter the Great in 1724. From its very beginning, it had a strong emphasis on fundamental research in science, engineering, and the humanities.

UNIVERSITY OF WÜRZBURG. This university is one of the oldest institutions of higher education in Germany and was founded in 1402. It was closed in 1415 but reopened in 1582. **Wilhelm Röntgen** was head of the Physics Department at the university, and this is where he investigated the properties of **cathode** rays and eventually discovered X-rays.

URANIUM. Uranium is a weakly radioactive element with the atomic number 92. All isotopes of uranium are unstable and have half-lives varying between 159,200 years and 4.5 billion years. In nature, it is found as uranium-238, uranium-235, and a small amount of uranium-234. U-235 is the only naturally occurring fissile isotope. Uranium was discovered in the mineral **pitchblende** by **Martin Heinrich Klaproth**, who named it after the recently discovered planet Uranus.

V

VASSAR COLLEGE. This college, founded in 1861 by Matthew Vassar, was the second degree-granting institution of higher education for **women** in the United States. It became coeducational in 1969. **Marie Curie** visited Vassar on her first trip to the **United States** in 1921 and addressed the students and faculty on 14 May 1921. This address was the only one made by Curie on this trip. Curie was very impressed by the three women's colleges that she visited (Smith, Mount Holyoke, and Vassar).

VILLARD, PAUL (1860–1934). Villard was a French chemist and physicist who discovered gamma rays in 1900 as he was studying the radiation emanating from **radium**. These rays were later called gamma rays and had a neutral charge. They were the third set of rays (alpha and beta) and made it apparent that the immutability of the atom was a myth.

VOGT, OSKAR (1870–1959). Oskar and **Cécile Vogt** represent another example of marital collaboration. They met in 1897 in Paris and married in 1899. They worked together throughout their lives and had a considerable influence on international neurological sciences. As a clinician, Vogt used hypnotism and was very interested in locating the origins of "genius" in the brain.

VOGT-MUGNIER, CÉCILE (1875–1962). Cécile Vogt-Mugnier was a neurologist who received a medical doctorate in Paris in 1900. Cécile and her husband, **Oskar Vogt**, were another example of a scientific couple, like the Curies, who worked together in close collaboration throughout the entirety of their careers. They met in Paris while Oskar was training with the neurologist and neuroanatomist Jules Dejerine, who had provided a model for this kind of marital collaboration with his own wife, Augusta Dejerine-Klumpke. The couple established a private neurobiological institute in Berlin and financed it with their neurological and psychiatric practices. As their fame grew, the private laboratory became associated with the University of Berlin, where Oskar Vogt taught. Many notable brain researchers were trained in their laboratories. The couple had two daughters, Marthe and Marguerite, who followed in their mother's footsteps and came to work in their parents' laboratory. The Kaiser Wilhelm Institute began to build a large complex of laboratories for the Vogts and their associates beginning in 1929 and completed in 1931. This brain institute included state-of-the-art electronics for recording brain impulses; laboratories for genetics, neuropathology, neurophysiology, neurochemistry; and even a sixty-bed neurological clinic. However, with the rise of the Nazi state, which they opposed, they moved to Neustadt in southern Germany and opened a private brain research institute. They continued to be active after World War II and began a study of aging in nerve cells. After Oskar's death, Cécile left Germany and went to live with her daughter, Marthe, in Cambridge, England, where the daughter had become a well-known neuropharmacologist.

W

WARSAW. The city of Warsaw, the capitol of **Poland**, located on the Vistula River in east-central Poland, was founded in the 13th century. Today, it is a dynamic metropolis and an important center of research and development. When **Marie Curie** was growing up, it was controlled by Russia and had been a victim of numerous partitions.

WOMEN. Although women have engaged in some aspect of the scientific enterprise from antiquity and throughout time, their numbers have been few. These few tended to be involved in the practical side of science, especially medicine. As the scope of science has changed throughout history, so has women's participation. The extent of women's involvement in science depends, in part, on society's views on gender roles. Throughout time, certain societies have been more accepting of women interested in the scientific endeavor than others, but generally science has been considered a masculine activity.

The idea of separate spheres for men and women, men occupying the public sphere and women the private, dominated social thought from the late 18th through the 19th century and spilled over into the 20th century. This theory partially explains the difficulty women who were interested in science encountered during **Marie Curie**'s time. By the late 19th century, in many parts of the world women could receive an advanced education but could not find a paying position. Marie Curie tenaciously prepared herself for a scientific career. Although her brilliance, creativity, and stubbornness were never in question, her success was an anomaly. Married women were seldom if ever appointed to a professorship at a world-renowned university as she was. She took advantage of one of the strategies that women scientists used, collaboration with a spouse to do her research. Until **Pierre Curie**'s death, she did not have a coveted position at the **Sorbonne**.

During the 19th and early 20th centuries, the higher education systems underwent upheavals in many parts of the world, especially the higher education of women. In the United States, the advent of the "Seven Sisters" women's colleges in the northeast—Vassar (1865), Smith (1875), Wellesley (1875), Radcliffe (1879), Bryn Mawr (1885), Barnard (1889), and Mount Holyoke (1893)—gave women the opportunity for higher education. In England, women established their own residential colleges—Girton (1869) and Newnham (1875) at Cambridge followed by Somerville (1879), Lady Margaret Hall (1879), St. Hugh's (1889), and St. Hilda's (1893) at Oxford. Although neither Oxford or Cambridge granted degrees to women during the 19th century, the examinations at these universities were gradually opened to them. The University of London and the provincial universities were more hospitable to women than Oxford and Cambridge. In Germany throughout the 19th century, women were unable to matriculate at universities although inroads were made toward the end of the century. Some women were allowed to attend as auditors, and some institutions granted degrees to foreign students. By the first decade of the 20th century, most of the legal barriers to women's admission crumbled. Because few German women had sufficient training to pass a matriculation examination, most of the women who entered German universities were foreigners; this situation led to the reform of German secondary education. In the third quarter of the 19th century, Switzerland, Sweden, and Denmark all opened their universities to women. Although Italian universities had accepted women students during the Middle Ages and Renaissance, they had closed their doors during the late 18th and early 19th centuries. Russia rejected a petition for female admission to the universities in 1867, so Russian women could only access higher education through an informal situation whereby they could participate in a combination of public lectures by cooperating professors and discussions in private homes. Since the **Poland** of Marie Curie was subject to Russian laws, her early higher education was of the informal Russian type.

WOMEN NOBEL PRIZE LAUREATES (1903–2018). **Marie Curie** was the first woman to win a Nobel Prize; in 1903, she shared that prize with **Pierre Curie** and **Henri Becquerel**. As of 2018, Nobel Prizes have been awarded to 853 men and 51 women. Marie Curie won it twice, the second time being in 1911, when she won the prize in chemistry. By 2018, five women have won the prize in chemistry, three in physics, 12 in physiology or medicine, and the rest in nonscientific fields such as the peace prize, the prize in literature, and the Nobel Memorial Prize in Economic Sciences. The women who received the prize in the sciences, in addition to Marie Curie, are as follows: **Irène Joliot-Curie**, chemistry (shared with **Frédéric Joliot-Curie**); Gerty Theresa Cori, physiology or medicine (shared with Carl Ferdinand Cori and Bernardo Houssay); Maria Goeppert-Mayer, physics (shared with J. Hans D. Jensen and Eugene Wigner); Dorothy Crowfoot Hodgkin, chemistry; Rosalyn Sussman Yalow, physiology or medicine (shared with Roger Guillemin and Andrew Schally); Barbara McClintock, physiology or medicine; Rita Levi-Montalcini, physiology or medicine (shared with Stanley Cohen); Gertrude B. Elion, physiology or medicine (shared with James W. Black and George H. Hitchings); Christiane Nüsslein-Volhard, physiology or medicine (shared with Edward B. Lewis and Eric F. Wieschaus); Linda B. Buck, physiology or medicine (shared with Richard Axel); Elizabeth Blackburn, physiology or medicine (shared with Jack W. Szostak); Carol W. Greider, physiology or medicine (shared with Jack W. Szostak); Ada E. Yonath, chemistry (shared with Venkatraman Ramakrishnan and Thomas A. Steitz); May Britt Moser, physiology or medicine (shared with Edvard Moser and John O'Keefe); Tu Youyou, physiology or medicine (shared with William C. Campbell and Satoshi Omura); Donna Strickland, physics (shared with Gerard Mourou and Arthur Ashkin); Frances Arnold, chemistry (shared with Gregory Winter and George Smith).

WOMEN SCIENTISTS IN MARIE CURIE'S LABORATORY. See GLEDITSCH, ELLEN; JOLIOT-CURIE, IRÈNE; LESLIE, MAY SYBIL; PEREY, MARGUERITE; RAMSTEDT, EVA JULIA AUGUSTA.

WORLD WAR I, EXPERIENCE OF I. JOLIOT-CURIE IN. To avoid the Germans, 17-year-old Irène was taken to the French countryside with her sister. She begged her mother to allow her to help in the field and suggested several alternatives, including being a Red Cross nurse, serving as a secretary, or even teaching. Marie, who was running 20 mobile field hospitals that she had established, tried to convince her to stay and take care of her little sister, Eve. Irène, however, was adamant, and Marie, who really needed her help, finally gave in. In order to be useful, Irène studied nursing and radiology. These hospitals had primitive X-ray equipment, which made it possible to assist doctors in locating shrapnel in wounded soldier. The equipment was crude, and both suffered large doses of radiation exposure. *See also* WORLD WAR I, EXPERIENCE OF M. CURIE IN.

WORLD WAR I, EXPERIENCE OF M. CURIE IN. Though **Marie Curie**'s new **Radium Institute** was completed on 31 July 1914, it was not used for research for over four years because of World War I. France began full war mobilization on 1 August 1914, after Germany declared war on Russia. Germany declared war on France on 3 August and invaded neutral Belgium. The men who worked in the laboratory were mobilized, leaving only Marie and one mechanic who had heart problems. The French government moved from Paris to Bordeaux, followed by many Parisians. Marie had to decide how best to protect both her family and her radium stored in the Rue Curie laboratory.

Shortly before mobilization, Marie sent 17-year-old Irène and nine-year-old Eve with a Polish housekeeper and Polish governess to the little fishing village of l'Arcouest in Brittany for a holiday. Marie's scientific friends the Perrins and Borels were there. The two girls had very different ideas about the coming war; Eve was upset but Irène was excited about the possibility. Since Marie was unable to get to the village, she instructed Irène to follow the Perrins' and Borels' suggestions. Although Irène rebelled, Marie recognized that they were in a safer place and insisted that they remain there.

A third child, **radium**, also needed Marie's protection. She took it from Paris to Bordeaux on the train in a heavy lead-protected bag. After she returned to Paris, she found that the Battle of the Marne had begun. She chose to remain in Paris at the institute, although she was concerned about the long separation from her daughters. Even though the French and English eventually won, it was at a terrible cost in the loss of lives. Although the French army finally retreated, it was still capable of fighting. Nevertheless, the victory made it possible for Irène and Eve to return to Paris and continue their schooling.

Marie was concerned not only about the events in France, she also was apprehensive about her family in **Poland**, partially occupied by the Germans. She made the decision to leave Paris for a war project that she was uniquely qualified to undertake, organizing radiology studies for military hospitals. By the beginning of the war, physicians had realized that knowing the exact location of the bullets would give the wounded a better chance to survive. Some of the large hospitals and a few mobile units had X-ray equipment, but

Marie Curie driving portable radiological unit during World War I

no units were available near the battle zones. Partially, this was due to the official view that X-rays would not be useful. Curie was able to convince a private organization, Patronage des Blesses, to give funds for her project. Still, the army blocked her at every stage. She collected all of the necessary apparatus from laboratories and in storage and trained volunteer helpers to work in several stations throughout France. Although these were helpful, they were insufficient, so Curie in collaboration with the Red Cross outfitted an ordinary touring car with the necessary equipment to form a mobile unit that when any hospitals near Paris called could respond.

Curie proved to be an excellent fundraiser and was able to establish or improve numerous radiological installations and to equip and give the army 20 radiologic cars. Irène begged her mother to allow her to help in the project. Although Marie objected strenuously she eventually succumbed, and Irène proved to be a worthy assistant. Although finding the equipment was not easy, finding trained radiologists was more of a problem. Insisting that an intelligent person with some idea about electrical machinery could be trained to operate the equipment, the army finally asked Curie to conduct a training course for technicians. Her solution was to train **women** since they were not directly involved in fighting to do radiological work. She wrote a book about her experiences during the war titled *La radiologie et la guerre* (Radiology and the War).

WORLD WAR II, EXPERIENCE OF I. JOLIOT-CURIE IN. Irène Joliot-Curie contracted tuberculosis during World War II and spent several years in Switzerland, leaving her husband and children in occupied France. She made several trips back to France during the war and several times was detained at the Swiss border. In 1944, she brought the children back to Switzerland with her after deciding France was too dangerous for them.

X

X-RAYS. *See* RÖNTGEN, WILHELM.

Z

ZORAWSKI FAMILY. This family employed Maria Sklodowska as a governess after she left her first disastrous governess position in **Warsaw**. The position paid well, but it required that 18-year-old Maria leave her home and travel 50 miles north to Szczuki, a place she had no knowledge of. When she first arrived, she was pleased with the family that had hired her. They had three sons in Warsaw, two in boarding school and one at the university. Living at home they had a daughter (Bronka), who was the same age as Maria; a 10-year-old girl (Andzia); a three-year-old boy (Stas); and a six-month-old baby girl (Maryshna). Because of the similarity in age, Maria found her relationship with Bronka confusing; as Bronka's tutor (three hours each day), she was considered Bronka's inferior, but they managed to become friends. Maria spent most of her tutoring time on Andzia (four hours a day) but found her an easily distracted student. After she finished her required seven hours of tutoring, she used her free time, with the Zorawskis' permission, teaching peasant children (from the Zorawskis' beet farms and the children from the sugar beet factories) how to read and write. Her first enthusiasm for her position was soon dampened. She found people of her own age, with the exception of Bronka, shallow and uninteresting. Although she was often bored, she used her spare time in self-education, reading books in physics, sociology, anatomy, and physiology. For further entertainment, she worked on problems in algebra and trigonometry. During this time, she decided that although she found many subjects interesting, she wanted a future in mathematics and physics. By the end of her stay, she no longer found the elder Zorawskis to be the excellent people that she had originally thought they were. The Zorawskis clearly believed that they were of a higher socioeconomic class than their employee, Maria, and objected to the romance of their son Kazimierz with Maria. *See also* ZORAWSKI, KAZIMIERZ.

ZORAWSKI, KAZIMIERZ (1866–1953). Kazimierz Zorawski fell in love with the family governess, Maria Sklodowsa, and she reciprocated his love. Although the two discussed marriage, Zorawski's parents rejected the match because of her family's poverty. Eventually, he accepted his parents' view of the marriage. Zorawski became a well-known mathematician, studying mathematics at the University of **Warsaw**, and continued his education in Leipzig and Göttingen, earning his Ph.D. degree in Leipzig for his thesis on the application of group conversion theory to differential geometry. In 1893, he received a doctorate in mathematics from Jagiellonian University in Krakow, where he was named assistant professor. He became full professor at Jagiellonian in 1898 and later a rector at that university. In 1919, he moved to Warsaw and became a full professor at the Warsaw University of Technology. He was elected to the Warsaw Society of Science and Letters and served as its president from 1926 to 1931. In 1926, he became a full professor of mathematics at the

University of Warsaw and was a Polish delegate for the International Committee on Intellectual Cooperation, which was established in 1922. **Marie Curie** was a prominent member of this organization, but there is no evidence that they ever met. He retired in 1935 and became professor emeritus in mathematics and natural science. In 1952, he was named a full member of the Polish Academy of Sciences. After his death, the importance of his work in Polish mathematics was recognized by scientists. He married a well-known pianist, Leokadia Jewniewicz, and the couple had three children. *See also* ZORAWSKI FAMILY.

Bibliography

CONTENTS

I. Primary Sources
   A. Archival Collections
   B. Archival Material
   C. Published Works
      a. Correspondence
      b. Conferences at Which Curie Discussed Her Research
      c. Editions and Translations of Curie's Doctoral Dissertation
      d. Publications during Curie's Lifetime
      e. Posthumous Publications
      f. Editions and Translations of Curie's Biography of Pierre Curie

II. Secondary Sources
   A. Curie in Context: Works That Situate Marie Curie in Place, Time, and Gender
   B. Scientific Predecessors and Colleagues
   C. Biographical Works That Emphasize Curie's Scientific Achievements
   D. Biographical Works That Emphasize Curie's Personal Life

INTRODUCTION

Although many books have been written about Marie Curie's life, her actual published output, while exceedingly important, was not as prolific as, for example, that of Charles Darwin. No complete bibliography of Curie's works exists. With few exceptions, her publications were first published in French; however, they were soon translated into other languages, including Polish, English, and German. Although most of the entries included in this bibliography are in English, it also includes selected works in the original language. The first section of this bibliography includes works by Curie herself, both unpublished and published. The unpublished source section, listed under "Archival Collections" and "Archival Material," does not purport to include all of the manuscript sources. The Curie archival material presents a unique barrier for the researcher since much of the material is radioactive even many years after her death. An investigator would find many of Curie's notes sealed in a lead-lined box in the French National library, the Bibliothèque Nationale. In order to examine these manuscripts, the researcher is required to sign a liability waiver. As letters by Curie emerged, they were soon printed. A sample of these published letters, both personal and professional, is included in this bibliography. The citations are arranged by publication date, not by the date the letters were originally written. Following the correspondence section, editions and translations of Curie's seminal work, her 1903 doctoral dissertation, are also arranged chronologically. Following this listing, other important publications by Curie are listed alphabetically.

Unlike the relatively modest number of Curie's actual publications, the secondary literature about her is immense. Undoubtedly, she is of great importance to the development of modern science. However, the fact that she was a woman successfully flouting the mores of her time regarding the traditional roles of wife and mother made her a prime subject for biographies. Her success in negotiating these diverse tasks suggested the possibility that a woman could be a highly successful scientist without abdicating traditional female responsibilities. Curie became a cultural icon. She is seen as that unusual person who successfully combined the roles of scientist, wife, and mother. When asked to name an important woman scientist, the first name that comes to mind is that of Marie Curie. She managed this difficult task through what Helena M. Pycior describes as her "anti-natural path." This path involved both Marie and Pierre's decision to concentrate only upon the essentials of life, which for them were science and their extended family. They were uninterested in traditional activities such as fashion. They had only sparse, utilitarian wardrobes, and they possessed only the necessary furniture. Their social ties were meager and seldom went beyond their extended families and a few selected scientists. They pared down family life to the barest essentials. After daughter Irène was born, several fortuitous events made it possible for Marie to continue with her scientific work. Pierre's father, Dr. Eugène Curie, took over much of the role of child care and, after the death of Pierre, made it possible for Marie to continue in the difficult position of an independent scientist and professor and single parent of, by that time, two little girls. These factors partially explain the great number of popular biographies of Curie. There are biographies of all kinds in many different languages, including scholarly biographies, popular biographies, and children's biographies.

The second main section of this bibliography lists secondary sources. Books and monographs, articles in periodicals, and chapters in books about Curie and her environment are interfiled in alphabetical order under subject headings. The first and second subdivisions place Curie's life and work in context, addressing the works of her scientific predecessors and contemporaries, world politics, women in science, and the state of university education. Biographies of Curie make up the greater part of this bibliography. They vary greatly, from beautifully researched productions to popular works, including children's biographies, coloring books, and photographs.

Although it is impossible to separate Curie's scientific life from her personal life, certain works stress her scientific achievements more than her personal life, whereas others emphasize the personal over the scientific. Recognizing that these two aspects of her life are intertwined and that dividing them is artificial, this bibliography nevertheless separates the works that stress her scientific life from those that emphasize her personal life. Curie's scientific works were a vital part of a revolution in physics and chemistry in the late 19th and early 20th centuries, and consequently those works that emphasize her scientific importance are listed first, although all of the entries note ways in which her science and personal life intersect. Because of Curie's unique position as a woman scientist with children, her way of managing home life and career is of interest to historians of women in science, and some of the biographies address this aspect of her career.

PRIMARY SOURCES

Archival Collections

Archives M. Curie at the Institut Curie, Paris.
Pierre and Marie Curie Papers, Bibliothèque Nationale, Paris. (Specific items are designated CP with their number.)
Institut Curie, Paris: Archives M. Curie. (Specific items are designated AC with their number.)

Archival Material

Chamié, Catherine. Five-page reminiscence *A la mémoire de Madame P. Curie*, 1935. AC, 433.
Correspondence with the company Sel de Radium, 1920–1929. AC, 1929.
Correspondence with the Société Géneral Métallurgique de Hoboken (Union Minière du Haute Katanga), 1923–1934.
Curie, Marie. *Exposé relative au radium qui se trouve actuellement dans mon laboratoire*, 3 March 1912. CP, n.a.f. 18436.
Frédéric Haudepin to Marie Curie, 3 February 1922. AC, 3146.

Note by Curie on visit to the factory, 29 April 1925. AC, 3142.

Notebooks and papers originally belonging to Marie Curie, Wellcome Collection, London.

Simcox, Lindsey, and Jon Taylor. *An Analysis of a Notebook and Papers, Originally Belonging to Marie Curie, Which Are Now Retained by the Wellcome Collection, London.* Aurora Health Physics Services, Harwell, Oxford, UK.

Published Works

Correspondence

The published and unpublished letters of Marie Curie are found in different locations and different languages.

Curie, Marie Sklodowska. *Correspondance: Choix de Lettres, 1905–1934 of Marie and Irène Curie.* Presented by Gilette Ziegler. Paris: Éditeurs Français Réunis, [1974].

———. "Listy Marii Sklodowskiej-Curie do Tatiany Jegorowej." (Letters of Marie Curie to Tatianie Jegorowej). Edited by Jerzy Róziewicz. *Kwartalnik Historii Nauki i Techniki* 25 (1980): 187–90.

———. "Materialy dotyczace Marii Sklodowskiej-Curie w archiwach Leningradu I Moskwy." (Materials on Marie Sklodowska-Curie in the archives of Leningrad and Moscow.) Edited by Jerzy Róziewicz. *Studia I Materialy z Dziejów Nauki Polskiej. Seria C: Historia Nauk Matematycznych, Fizykochemicznych I Geoloiczno-geograficznych* 25 (1981): 77–82.

Tešinská, Emillie. "Z zachowanej w Checho-slowacji kspondenscji Marii Sklodowskiej-Curie." (On the Curie Correspondence Recently Discovered in Czechoslovakia.) *Kwartalnik Historii Nauki i Techniki* 33, no. 4 (1988): 509–22.

Curie, Marie Sklodowska. *Korespondencja polska Marii Skłodowskiej-Curie, 1881–1934, Opracowanie, Kabzińska, Krystybam et al.* Instytut Historii Nauki PAN Polski Towarzystwo Chemicznem, 1994.

———. *Korespondencja Marii Skłodowskiej-Curie z uczonymi z Europy Środkowej i Wschodniej, 1904–1934.* Edited by Jana Piskurewicza. Lubin: Wydawn. Uniwersytetu Marii Curie-Sklodowskiej, 1998.

———. *Lettres: Marie Curie et ses filles.* Edited by Hélène Langevin-Joliot and Monique Bordry. Paris: Pygmalion, 2011.

———. *Listy Maria Skłodowska-Curie.* Warszawa: Drzewo Babel, Alta Studio, Alicja Albrecht, 2012.

Conferences at Which Curie Discussed Her Research

Conférence faite le 7 Mars 1920 sur les Radio-éléments et leurs Applications. Paris: Librairie de l'Eseignement Technique, 1920.

Memorandum by Marie Curie, member of the Committee on the Question of International Scholarships for the Advancement of the Sciences and the Development of Laboratories. League of Nations. International Committee on Intellectual Co-operation. Geneva: d'Ambilly, [1926].

Editions and Translations of Curie's Doctoral Dissertation

Marie Curie's doctoral thesis was approved on 11 May 1903, by the Sorbonne's dean of the faculty, Paul Appell. In June 1903, she defended it. Recognized as an important document in the science of radioactivity, it was first published in England over the summer in Crooke's *Chemical News* and in France in the *Annales de physique et de chimie* in September 1903. Subsequently, it was reprinted and translated many times. Editions in chronological order.

Curie, Marie. *Recherches sur les substances radioactives.* Paris: Gauthier-Villars, 1903.

———. *Radio-Active Substances: Thesis Presented to the Faculté des Sciences de Paris.* London: Chemical News Office, 1904.

———. *Radioactive Substances: A Translation from the French of the Classical Thesis Presented to the Faculty of Science in Paris by the Distinguished Nobel Prize Winner Marie Curie.* New York: Philosophical Library, 1961.

———. *Radioactive Substances: A Translation from the French of Curie's Thesis Presented to the Faculty of Sciences in Paris.* Westport, CT: Greenwood Press, 1971.

———. *Radioactive Substances.* Mineola, NY: Dover, 2002.

Publications during Curie's Lifetime

Curie, Marie. "Eléctromètre à fil de quartz." *Le radium*, no. 3 (1906): 202–3.

———. *La radiologie et la guerre.* Paris: Gauthier-Villars, 1921.

———. "Les mesures en radioactivité et l'étalon du radium." *Journal de Physique* 2 (1912): 715–826.

———. "Les Rayons de Becquerel et le polonium." *Revue Général des Sciences* 10 (1899): 41.

———. *L'Isotopie et les éléments isotopes.* Paris: Librairie Scientifique Albert Blanchard, 1924.

———. Note presented to the *Academia* by M. Lippmann. "Rays Emitted by Compounds of Uranium and of Thorium." *Comptes Rendus* 126 (1898): 1101–3.

———. *Radioactivité.* Paris: Hermann, 1935.

———. *Radium and Radioactivity.* New York: Century Magazine, 1904.

———. *Rayons [Alpha], [Beta], [Gamma] des corps radioactifs en relation avec la structure nucléaire.* Paris: Hermann, 1933.

———. "Rayons émis par les composes de l'uranium et du thorium." *Comptes Rendus* 126 (April 1898): 1101–3.

———. *Recherches sur les substances radioactives.* Paris: Gauthier-Villars, 1903.

———. "Sur l'actinium. *Journal de Chimie et Physique* 27 (1930): 1.

———. "Sur le poids atomique du radium." *Comptes Rendus de l'Académie des Sciences* 145 (1907): 422.

———. *Traité de radioactivité.* 2 vols. Paris: Gauthier-Villars, 1910.

———. Über den radioaktiven Stoff 'Polonium.'" *Physikalische* Zeitschrift 4 (1903): 234.

Curie, Marie, and Pierre Curie. Note presented by M. Becquerel. "On a New Radioactive Substance Contained in Pitchbende. *Comptes Rendus* 127 (1898): 175-178.

Curie, Pierre, Marie Skłodowska Curie, and Gustave Bémont. "On a New 'Strongly' Radioactive Substance Contained in Pitchblende." *Comptes Rendus de l'Academie des Sciences* 127 (1898): 1215–17.

Posthumous Publications

Boorse, Henry A., and Lloyd Motz, eds. *The World of the Atom.* 2 vols. New York: Basic Books, 1966. Includes papers by Curie.

Curie, Marie. *Entdeckung des Radiums: Rede gehalten am 11 Dezember 1911 in Stockholm beim Empfang des Nobelpreises für Chemie; Untersuchungen über die radioaktiven Substanzen.* Thun, Switzerland: H. Deutsch, 1999.

———. *Oeuvres de Marie Skłodowska Curie.* Warsaw: Państwowe Wydawnictwo Naukowe, 1954

———. *Prace Marii Skłodowskiej-Curie, zebrane przez Irène Joliot-Curie.* Warsaw: Państwowe Wydawn, 1954.

———. *Selbestbiographie.* Translated from the Polish. Leipzig: B.G. Teubner, 1962.

Editions and Translations of Curie's Biography of Pierre Curie

Marie Curie's biography of Pierre has been reprinted in many different forms. Some contain Marie's autobiographical notes and others do not.

Marie Curie. *Pierre Curie: With Autobiographical Notes.* New York: Macmillan, 1923.

———. *Pierre Curie.* Paris: Payot, 1924.

———. *Pierre Curie: Avec une études des "Carnets de laboratoire" par Irène Joliot-Curie.* Warsaw: Państwowe Wydawn Naukowe, 1954.

SECONDARY SOURCES

Curie in Context: Works That Situate Marie Curie in Place, Time, and Gender

In order to situate Marie Curie's work in her environment, books that describe her physical and intellectual surroundings are listed separately from those that concentrate on her life and work. Descriptions of the Poland of her early years, the Paris of her creative years, her influence on radioactivity in different countries, and the influence of other scientists on her achievements are listed in the following section. Although women as creative scientists certainly existed before and during Marie Curie's lifetime, she was unique in her success as a woman scientist. As a woman and a scientist, Curie must also be situated in a gendered society. Books concerned with the position of women in society, especially those women involved in science, are included in this section.

Abbot, Mary. *A Woman's Paris: A Handbook of Every-Day Living in the French Capitol.* Boston: Small, Maynard & Co., 1900.

Abir-Am, Pnina G., and Dorinda Outram. "Mme Curie's 2011 Centennial and the Public Debate on the Underrepresentation of Women in Science: Lessons from the History of Science." In *Celebrating the 100th Anniversary of Madame Marie Sklodowska Curie's Nobel Prize in Chemistry*, edited by M. H. Chiu, P. J. Gilmer, and D. F. Treagust, 205–23. Rotterdam: Sense, 2011.

———, eds. *Uneasy Careers and Intimate Lives: Women in Science, 1789–1979.* New Brunswick, NJ: Rutgers University Press, 1987.

Adloff, Jean-Pierre, and George B. Kauffman. "Marguerite Perey (1909–1975): A Personal Retrospective Tribute on the 30th Anniversary of Her Death." *Chemical Educator* 10, no. 5 (2005): 378–86.

Ajalbert, Jean. *Dans Paris, la grande ville (sensations de guerre).* Paris: Éditions Georges Crès, 1916.

Armet de isle, Émile. "Électrosope pour la recherche des minéraux radioactifs." *Le Radium* 3 (1906): 62.

Badash, Lawrence. "Radium, Radioactivity, and the Popularity of Scientific Discovery." *Proceedings of the American Philosophical Society* 122 (June 1978): 145–54.

Baedeker, K. *Paris and Environs.* Leipzig: Karl Baedeker, 1891.

Barbusse, Henri. *Le Feu.* Paris: Flammarion, 1965. A novel about trench warfare, first published in French in 1916.

Becker, Jean-Jacques. *The Great War and the French People.* New York: St. Martin's Press, 1986.

Billinski, Lucjan. *Z Mazowsza do slawy paryskiego Panteonu.* (*On Marie Curie and Her Polish Heritage*). Warsaw: Biblioteka Główna Województwa Mazowieckego, 2003. In Polish but includes a summary in English.

Blejwas, Stanislaus Andrew. "Warsaw Positivism: 1864–1890: Organic Work as an Expression of National Survival in Nineteenth-Century Poland." Ph.D. diss., Columbia University, 1973.

Bourrelier, Henri. *La Vie du quartier-latin.* Paris: Éditions Bourrelier, 1936.

Boyd, Louise. *Polish Countrysides.* New York: American Geographic Society, 1837.

Brandes, George. *Poland: A Study of the Land, People and Literature.* New York: Macmillan, 1903.

Bromke, Adam. *Poland's Politics: Idealism versus Realism.* Cambridge, MA: Harvard University Press, 1967.

Caullery, Maurice. *French Science, and Its Principal Discoveries.* New York: Harper and Row, 1975.

Charle, Christophe, and Eva Telkes. *Les Professeurs de la Faculté des sciences de Paris: Dictionnaire biographique 1901–1939.* Paris: Editions du CNRS, 1989.

Charrier, Edemée. *L'Évolution intellectuelle feminine: Thèse pour le doctorate en droit.* Paris: Editions Albert Mechelinck, 1931.

Clark, C. *Radium Girls: Women and Industrial Health Reform, 1910–1935.* Chapel Hill: University of North Carolina Press, 1997.

Cobban, Alfred. *A History of Modern France.* Reading, UK: Cox and Wyman, 1965.

Crawford, Elisabeth. *The Beginnings of the Nobel Institution: The Science Prizes, 1901–*

1915. Cambridge: Cambridge University Press, 1984.

———. "The Prize System of the Academy of Sciences." In *The Organization of Science and Technology in France, 1808–1914*, edited by Robert Fox and George Weisz, 283–307. Cambridge: Cambridge University Press, 2009.

Crawford, Elisabeth, J. L. Heilbron, and Rebecca Urich. *The Nobel Population 1901–1937*. Berkeley, CA: Office for History of Science and Technology, 1987.

Des Jardins, Julie. "American Memories of Madame Curie: Prisms on the Gendered Culture of Science." In *Celebrating the 100th Anniversary of Madame Marie Sklodowska Curie's Nobel Prize in Chemistry*, edited by M. H. Chiu, P. J. Gilmer, and D. F. Treagust, 59–85. Rotterdam: Sense, 2011.

———. *American Queenmaker: How Missy Meloney Brought Women into Politics*. New York: Basic Books, 2019.

Elena, Alberto. "Skirts in the Lab: Madame Curie and the Image of the Woman Scientist in the Feature Film." *Public Understanding of Science* 6 (1997): 269–78.

Fara, Patricia. "Curicatures." *Endeavour: Review of the Progress of Science* 28 (2004): 101–3.

Fellinger, Anne. "Women Radio-Chemists Facing Radioactive Risks in France." In *The Global and the Local: The History of Science and the Cultural Integration of Europe, Proceedings of the 2nd ICESHS*, 534–539. Krakow: ICESHS, 2006.

Fox, Robert, and George Weicz, eds. *The Organization of Science and Technology in France 1808–1914*. Cambridge: Cambridge University Press, 1980.

Fredrickson, Anne. "Vanity Unfair." *Distillations*, 16 June 2011. https://www.sciencehistory.org/distillations/vanity-unfair.

Gablot, Ginette. "A Parisian Walk along the Landmarks of the Discovery of Radioactivity." *Physics in Perspective* 2 (2000): 100–107.

Gilbert, John K. "Women Chemists Informing Public Education about Chemistry during the 20th Century." In *Celebrating the 100th Anniversary of Madame Marie Sklodowska Curie's Nobel Prize in Chemistry*, edited by M. H. Chiu, P. J. Gilmer, and D. F. Treagust, 141–66. Rotterdam: Sense, 2011.

Harding, Warren G. *Remarks of the President in Presenting to Madam Curie a Gift of Radium from the American People*. Washington, DC: Government Printing Office, 1921.

Hause, Steven C., with Anne R. Kenney. *Women's Suffrage and Social Politics in the French Third Republic*. Princeton, NJ: Princeton University Press, 1984.

Heilbron, J. L. "Fin-de-siècle Physics." In *Science, Technology and Society in the Time of Alfred Nobel*, edited by Carl Gustaf Bernhard, Elisabeth Crawford, and Per Sörbom, 51–73. Oxford: Pergamon Press, 1982.

Jaquemond, Louis Pascal, *Irène Joliot-Curie*. Paris: Odile Jacob, 2014.

Kawashima, Keiko. "Female Scientists Whom Nobuo Yamada Encountered: Early Years of Radio Chemistry and the Radium Institute." In *Proceedings of the International Workshop on the History of Chemistry: Transformation of Chemistry from the 1920s to the 1960s*, 56–67. Tokyo: IWHC, 2015.

Mamlok-Naaman, Rachel, Ron Blonder, and Yehudit Judy Dori. "One Hundred Years of Women in Chemistry in the 20th Century: Sociocultural Developments of Women's Status." In *Celebrating the 100th Anniversary of Madame Marie Sklodowska Curie's Nobel Prize in Chemistry*, edited by M. H. Chiu, P. J. Gilmer, and D. F. Treagust, 119–39. Rotterdam: Sense, 2011.

McAuliffe, Mary Sperling. *Twilight of the Belle Epoque: The Paris of Picasso, Stravinsky, Proust, Renault, Marie Curie, Gertrude Stein, and Their Friends through the Great War*. Lanham, MD: Rowman & Littlefield, 2014.

McGrayne, Sharon Bertsch. *Nobel Prize Women in Science: Their Lives, Struggles, and Momentous Discoveries*. New York: Birch Lane Press, 1993.

McMillan, James F. *Twentieth-Century France: Politics and Society 1898–1991*. London: Edward Arnold, 1992.

Montorgueil, Gorges. *Les Parisiennes d'à present*. Paris: H. Floury, 1897.

Mould, Sharon Bertsch. *Nobel Prize Women in Science: Their Lives, Struggles, and*

Momentous Discoveries. New York: Carol Press, 1998.

Ogilvie, Marilyn Bailey, and Joy Harvey, eds. *The Biographical Dictionary of Women in Science: Pioneering Lives from Ancient Times to the Mid-20th Century*. 2 vols. New York: Routledge, 2000.

———. "Marie Curie, Women and the History of Chemistry." In *Celebrating the 100th Anniversary of Madame Marie Sklodowska Curie's Nobel Prize in Chemistry*, edited by M. H. Chiu, P. J. Gilmer, and D. F. Treagust, 105–18. Rotterdam: Sense, 2011.

Palmer, William P. "Forgotten Women in Science Education: The Case of Mary Amelia Swift." In *Celebrating the 100th Anniversary of Madame Marie Sklodowska Curie's Nobel Prize in Chemistry*, edited by M. H. Chiu, P. J. Gilmer, and D. F. Treagust, 167–87. Rotterdam: Sense, 2011.

Paul, Harry W. *From Knowledge to Power: The Rise of the French Science Empire in France, 1860–1939*. Cambridge: Cambridge University Press, 1985.

Paxton, Robert. *Europe in the Twentieth Century*. New York: Harcourt Brace, 1985.

Pigeard-Micault, Natalie. *Femmes du laboratoire de Marie Curie*. Paris: Éditions Glyphe, 2013.

Rayner-Canham, Marelene F., and Geoffrey W. Rayner-Canham, eds. *A Devotion to Their Science: Pioneer Women of Radioactivity*. Philadelphia: Chemical Heritage Foundation, 1997.

Roqué, Xavier. "Ciencis e industria en el desararollo de la radiactividad: El caso de Marie Curie." *Arbor: Ciencia, Pensamiento y Cultura* 156 (1996): 25–49.

Rossiter, Margaret W. *Women Scientists in America: Struggles and Strategies to 1940*. Baltimore: Johns Hopkins University Press, 1982.

Róziewicz, Jerzy. "Zwiatzki Marii Sklodowskiej-Curie z nauka rosyjska I radziecka" (Contacts of Marie Curie with Russian and Soviet Science). *Kwartalnik Historii Nauki i Techniki* 29 (1984): 535–55.

Schürmann, Astrid. "Marie Curie und ihr Laboratoire: Frauenförderung avant la lettre." *Finska Feministische Studien* 24 (2006): 29–45.

Sharp, Evelyn. *Hertha Ayrton: A Memoir*. London: Edward Arnold, 1926.

Southerland, Sherry A., and Sibel Uysal Bahbah. "Educational Policy of Accountability and Women's Representation in Science: The Specter of Unintended Consequences." In *Celebrating the 100th Anniversary of Madame Marie Sklodowska Curie's Nobel Prize in Chemistry*, edited by M. H. Chiu, P. J. Gilmer, and D. F. Treagust, 225–38. Rotterdam: Sense, 2011.

Stetson, Dorothy McBride. *Women's Rights in France*. New York: Greenwood Press, 1987.

Strohmaier, Brigitte, and Robert Rosner. *Marietta Blau, Stars of Disintegration: Biography of a Pioneer of Particle Physics*. Riverside, CA: Ariadne Press, 2013.

Thébaud, Françoise. *La Femme au temps de la guerre de 1*. Paris: Éditions Stock, 1962.

Weicz, George. *The Emergence of Modern Universities in France, 1863–1914*. Princeton, NJ: Princeton University Press, 1968.

Zamoyski, Adam. *The Polish Way: A Thousand-Year History of the Poles and Their Culture*: New York: Franklin Watts, 1988.

Scientific Predecessors and Colleagues

Marie Curie's vital scientific work in radioactivity had many predecessors in chemistry and physics. They are a part of her bibliography, as are those who built their reputations on her work. This list also includes selected works of her contemporaries.

Badash, Lawrence. "Decay of a Radioactive Halo." *Isis* 66, no. 4 (1975): 566–68.

———. "Radioactivity before the Curies." *American Journal of Physics* 33, no. 2, (1965): 128–35.

———. *Radioactivity in America: Growth and Decay of a Science*. Baltimore: Johns Hopkins University Press, 1979.

———. "Radium, Radioactivity, and the Popularity of Scientific Discovery." *Proceedings of the American Philosophical Society* 122 (June 1978): 145–54.

———. *Rutherford and Boltwood: Letters on Radioactivity*. New Haven, CT: Yale University Press, 1969.

Bensaude-Vincent, Bernadette. *Langevin: Science et vigilance*. Paris: Belin, 1987.

———. "Une robe de cotton noir." *Cahiers de Science & Vie* 24 (1994): 76–85.

Bensaude-Vincent, Bernadette, and Isabelle Stengers. *Histoire de la chimie*. Paris: Éditions la Découverte, 1993.

Biezunski, Michel. *Albert Einstein: Correspondances françaises*. Paris: Éditions de Seuil, 1989.

Bimbot, René. *Histoire de la Radioactivité: l'évolution d'un concept et de ses applications*. Paris: Vuibert, 2006.

Boorse, Henry A., and Lloyd Moltz, eds. *The World of the Atom*. Vol. 1. New York: Basic Books, 1996. Includes translations of Curie papers.

Boudia, Soraya. "The Curie Laboratory: Radioactivity and Metrology." *History and Technology* 13, no. 4 (1997): 249–65.

Boudia, Soraya, and Xavier Roqué, eds. "Science, Medicine, and Industry: The Curie and Joliot-Curie Laboratories." Special issue, *History and Technology* 13 (no. 4, 1997): 241–343.

Caulfield, Catherine. *Multiple Exposures: Chronicles of the Radiation Age*. New York: Harper & Row, 1989.

Chiu, M. H., P. J. Gilmer, and D. F. Treagust, eds. *Celebrating the 100th Anniversary of Madame Marie Sklodowska Curie's Nobel Prize in Chemistry*. Rotterdam: Sense, 2011.

Cosset, J. M. *Fantastique histoire du radium: quand un élément radioactive deviant potion magique*. Rennes: Ouest-France, 2011.

Cregan, Elizabeth R. *Marie Curie: Pioneering Physicist*. Minneapolis: Compass Point Books, 2007.

Crossland, Maurice. *Science under Control*. Cambridge: Cambridge University Press, 1992.

Curie, Pierre. *Oeuvres de Pierre Curie*. Paris: Gauthier-Villars, 1908.

Demarcay, Eugène. "Sur lle spectre d'une substance radio-active." *Comptes Rendus* 127 (1898): 1218.

Eve, A. S. *Rutherford: Being the Life and Letters of the Rt Hon. Lord Rutherford, O.M.* New York: Macmillan, 1939.

Feather, Norman. "The Experimental Discovery of the Neutron." *Proceedings of the Tenth International Congress of the History of Science*. Paris: Academie Internationale d'Histoire Des Sciences, 1964.

Gilmer, Penny J. "Iréne Joliot Curie, a Nobel Laureate in Artificial Radioactivity." In *Celebrating the 100th Anniversary of Madame Marie Sklodowska Curie's Nobel Prize in Chemistry*, edited by M. H. Chiu, P. J. Gilmer, and D. F. Treagust, 41–57. Rotterdam: Sense, 2011.

Hussenius, Anita, and Kathryn Scantlebury. "Witches, Alchemists, Poisoners and Scientists: Changing Image of Chemistry." In *Celebrating the 100th Anniversary of Madame Marie Sklodowska Curie's Nobel Prize in Chemistry*, edited by M. H. Chiu, P. J. Gilmer, and D. F. Treagust, 191–204. Rotterdam: Sense, 2011.

Joerges, Bernward, and Terry Shinn, eds. *Instrumentation between Science, State and Industry*. Dordrecht: Kluwer Academic, 2001.

Jorgensen, Timothy J. *Strange Glow: The Story of Radiation*. Princeton, NJ: Princeton University Press, 2016.

Kawashima, Keiko. "D'une sainte à une femme ambitieuse: La transformation de l'image de Marie Curie et les biographies des femmes de science au XXe siècle." *Kagakushi* 28 (2001): 145–62.

Kevles, Daniel J. *The Physicists*. New York: Alfred A. Knopf, 1978.

Langevin, André. *Paul Langevin mon père*. Paris: Les Éditeurs Français Réunis, 1971.

Majumdar, Sisir K. "International Chemistry Year: Centenary of Marie Curie's Second Nobel Laurel." *Indian Journal of History of Science* 47, no. 2 (2012): 287–90.

Malley, Marjorie. "The Discovery of Atomic Transmutation: Scientific Styles and Philosophies in France and Britain." *Isis* 70 (1979): 213–23.

Mehra, Jagdish. *The Solvay Conferences on Physics: Aspects of the Development of Physics since 1911*. Boston: D. Reidel, 1975.

Molinié, Phillippe, and Soraya Boudia. "Exhibiting Sparks of Big Science to the Public:

Electrostatics, Atomic Machines and Experience of Paris Palais de la Découverte." *IEEE Transactions on Dialectrics and Electrical Insulation* 16, no. 3 (June 2009): 751–57.

Nye, Mary Jo. "Gustav LeBon's Black Light." *Historical Studies in the Physical Sciences* 4 (1974): 163–95.

———. *Molecular Reality: Perspective on the Scientific Work of Jean Perrin.* New York: American Elsevier, 1972.

———. Review of Barbo Loïc, *Pierre Curie (1859–1906): Le rêve scientifique. Isis* 92, no. 4 (December 2001): 789–90.

Ogilvie, Marilyn Bailey. "Marie Curie and Her Fellow Scientists." *Physics Today*, 7 November 2017.

Owens, Trevor. "Madame Curie above the Fold: Divergent Perspectives on Curie's Visit to the United States in the American Press." *Science Communication* 33 (September 2011): 98–119.

Pais, Abraham. *Inward Bound: Of Matter and Forces in the Physical World.* New York: Oxford University Press, 1986.

———. *Subtle Is the Lord.* New York: Oxford University Press, 1982.

Pestre, Dominique. "The Moral and Political Economy of French Scientists in the First Half of the XXth Century." *History and Technology* 13, no. 4 (1997): 241–48.

———. *Physique et physiciens en France, 1918–1940.* Paris: Éditions des Archives Contemporaines, 1984.

Pinault, Michel. Review of Barbo Loïc, *Pierre Curie (1859–1906): Le rêve scientifique. Vingtième Siècle. Revue d'histoire* 67 (2000): 209–10

Piskunowicz, J. "The Collaboration between Albert Einstein and Maria Sklodowska-Curie." *Kwartalnik Historii Nauki i Techniki* 50, nos. 3 & 4 (2005): 7–24.

Preston, Diana. *Before the Fallout: From Marie Curie to Hiroshima.* New York: Walker & Company, 2005.

Rayner-Canham, Geoff, and Zheng Zheng. "Naming Elements after Scientists: An Account of a Controversy." *Foundations of Chemistry* 10, no. 1 (2008): 13–18.

Schmidt, Gerhard Carl. "Über die von den Thorverbindungen und einigen anderen Substanzen ausgehende Strahlung." *Annalen der Physik und Chemie* 65 (1898): 141–51.

Weart, Spencer R. *Scientists in Power.* Cambridge, MA: Harvard University Press, 1979.

Whitlock, Catherine, and Evans Rhodri. *Ten Women Who Changed Science and the World: Marie Curie, Rita Levi-Montalicini, Chien-Schiung Wu, Virginia Apgar, and More.* New York: Diversion Press, 2019. Short biographies of 10 women scientists, including Marie Curie.

Biographical Works That Emphasize Curie's Scientific Achievements

Many of the works on Marie Curie are concerned both with her biographical details and her scientific work. However, they often emphasize one of these aspects of her life. The following citations, although they may consider biographical information, stress her scientific work.

Adloff, Jean-Pierre. "The Laboratory Notebooks of Pierre and Marie Curie and the Discovery of Polonium and Radium." *Czechoslovak Journal of Physics* 49 (1999): Suppl. S1.

Adloff, Jean-Pierre, and George B. Kauffman. "Pierre Curie (1859–1906): A Retrospective View on the Centenary of His Death (2006)." *Chemical Educator* 11, no. 2 (2006): 110–17.

Birch, Beverley. *Marie Curie: Pioneer in the Study of Radiation.* Milwaukee: G. Stevens Children's Books, 1990.

———. *Marie Curie: The Polish Scientist Who Discovered Radium and Its Life-Saving Properties.* Milwaukee: G. Stevens, 1988.

———. *Marie Curie's Search for Radium.* Hauppauge, NY: Barron's, 1996.

Blanc, Karin. *Marie Curie et le Nobel.* Uppsala: Uppsala University, Office for the History of Science, 1999.

Bodden, Valerie. *Marie Curie: Chemist and Physicist.* Minneapolis: Essential Library, 2017.

Bordry, Monique, and Soraya Boudia, eds. *Les rayons de la vie: Une histoire des applications médicales des rayons X et de la radioactivité en France, 1895–1930*. Paris: Institut Curie, 1998.

Borzendowski, Janice. *Marie Curie: Mother of Modern Physics*. New York: Sterling, 2009.

Boudia, Soraya. "Marie Curie et son laboratoire: science, industrie, instruments et métrologie de la radioactivité en France, 1896–1914." Ph.D. diss., University of Paris, 1997.

———. *Marie Curie et son laboratoire: sciences et industrie de la radioactivité en France*. Paris: Editions des Archives Contemporaines, 2001.

Brandt, Keith. *Marie Curie, Brave Scientist*. Mahwah, NJ: Troll Associates, 1983.

Burby, Liza N. *Marie Curie: Nobel Prize-Winning Physicist*. New York: Powerkids Press, 1997.

Chiu, Mei-Hung, and Nadia Y. Wang. "Marie Curie and Science Education." In *Celebrating the 100th Anniversary of Madame Marie Sklodowska Curie's Nobel Prize in Chemistry*, edited by M. H. Chiu, P. J. Gilmer, and D. F. Treagust, 9–39. Rotterdam: Sense, 2011.

Davis, J. L. "The Research School of Marie Curie in the Paris Faculty, 1907–14." *Annals of Science: The History of Science and Technology* 52, no. 4 (1995): 321–55.

Dickmann, Nancy. *Marie Curie: The Woman behind Radioactivity*. North Mankato, MN: Pebble, 2019.

Discorsi pronunciati dal prof. Józef Hurwic e dal lanceo Gilberto Bernardini della seduta del 10 Febbraio 1968. Rome: Accademia Nazionale del Lincei, 1968.

Ethridge, Maggie May. *Marie Curie: Radioactive Pioneer and First Woman to Win a Nobel Prize*. New York: Cavendish Square, 2017.

Fölsing, Ulla. *Marie Curie: Wegbe reiterin einer neyen Naturwissenschaft*. Munich: Piper, 1990.

Gooday, Graeme. "Domesticating the Magnet: Secularity, Secrecy and 'Permanency' as Epistemic Boundaries in Marie Curie's Early Work." *Spontaneous Generations* 3, no. 1 (2009): 68–81.

Graham, Ian. *Curie and the Science of Radioactivity*. Hauppauge, NY: Barons Educational Series, 1906.

Greene, Carol. *Marie Curie, Pioneer Physicist*. Chicago: Children's Press, 1964.

Hasday, Judy L. *Marie Curie: Pioneer on the Frontier of Radioactivity*. Berkeley Heights, NJ: Enslow, 2004.

Hellman, Samuel. "On the Discovery of Radioactivity by the Curies and Its Medical and Cultural Ramifications." *Perspectives in Biology and Medicine* 36 (1992): 39–45.

Hughes, Jeff. "The French Connection: The Joliot-Curies and Nuclear Research in Paris, 1925–1933." *History and Technology* 13, no. 4 (1997): 325–43.

Hurwic, Józef. "Importance de la thèse doctorat de Marie Skłodowska Curie pour le développement de la science sur la radioactivité." Thesis, Faculté des Sciences, University of Paris, 1992.

———. *Maria Skłodowska Curie*. Warsaw: Polonia, 1967.

Joliot-Curie, Irène. "Les Carnets de laboratoire de la découverte du polonium et du radium." An appendix to the French edition of Marie Curie's *Pierre Curie*, 1955 edition.

———. "Marie Curie, ma mère." *Europe* 108 (1954): 89–121.

Kabzińska, Krystyna. "Laboratorium Curie jako miedzynarodowy ośrodek ksztalcenia kadry naukowej i jego Portugalscy wychowankowie w latach 1914–1938" (The Curie Laboratory as an International Training Center of Scientific Cadres and Its Portuguese Pupils.) *Kwartalnik Hisstorii Nauki i Techniki* 33 (1988): 169–92.

Kawashima, Keiko. "Deux savants japoais et la famille Curie, Nobuo Yamada et Toshiko Yuasa." *L'Actualité Chimique* 363 (May 2012): 51–55.

Klickstein, Herbert S. *Maria Skłodowska Curie, Recherches sur les substances radioactives: A Bio-bibliographical Study*. St. Louis: Mallinckrodt Chemical Works, 1966.

Koestler-Grack, Rachel A. *Marie Curie: Scientist*. New York: Chelsea House, 2009.

Krieg, Katherine. *Marie Curie: Physics and Chemistry Pioneer*. Minneapolis: Core Library, 2015.

Kuleff, I. "Marie Sklodowska-Curie and the International Year of Chemistry." *Khimiya/Chemistry: Bulgarian Journal of Chemical Education* 20 (1911): 83–95.

Massinot, Anaïs, and Natalie Pigeard-Micault. *Marie Curie et la grande guerre*. Paris: Edition Glyphe, 2014.

McClafferty, Carla Killough. *Something out of Nothing: Marie Curie and Radium*. New York: Farrar, Straus and Giroux, 2006.

Mierzycki, Roman. "Dokumenty doktoratu Marii Sklodowskiej-Curie" (Documents on the Doctorala Promotion of M. Curie). *Kwartalnik Historii Nauki i Techniki* 37, no. 3 (1992): 101–10.

———. "Echa odkrycia polonu i radu w Polsce, ogólnodostepne wyklady I prasa w latach 1898–1901" (The Discovery of Polonium and Radium in the Polish Press, 1898–1901.) *Analecta: Studia I Materialy z Dziejów Nauki* 7, no. 1 (1998): 7–28.

Parker, Steve. *Marie Curie and Radium*. Philadelphia: Chelsea House, 1995.

Pasachoff, Naomi E. *Marie Curie and the Science of Radioactivity*. New York: Oxford University Press, 1996.

Perrin, Jean. "Madame Curie et la découverte du Radium." *Vient de paraitre, Bulletin bibiographique mensuel*, February 1925.

Pigeard-Micault, Natalie. "The Curie's Laboratory and Its Women (1906–1934)." *Annals of Science* 70 (2013): 71–100.

Poirier, Jean-Pierre. *Marie Curie et les conquérants de l'atome: 1896–2006*. Paris: Pygmalion, 2006.

Pycior, Helena M. "Reaping the Benefits of Collaboration While Avoiding Its Pitfalls: Marie Curie's Rise to Scientific Prominence." *Social Studies of Science* 23 (1993): 301–23.

Robison, Roger Frank. *Mining and Selling Radium and Uranium*. Heidelberg: Springer International, 2015.

Romer, Alfred, ed. *The Curies: Polonium and Radium, in Radiochemistry and the Discovery of Isotopes*. New York: Dover, 1970.

———, ed. *The Discovery of Radioactivity and Transmutation*. New York: Dover, 1964.

Roqué, Xavier. "The Curie Laboratories and the Radium Industry." In *Instrumentation between Science, State and Industry: Sociology of the Sciences*, edited by Bernward Joerges and Terry Shinn. Vol. 22. Dordrecht: Springer, 2001.

———. "Displacing Radioactivity." In *Instrumentation between Science, State and Industry: Sociology of the Sciences*, edited by Bernward Joerges and Terry Shinn. Vol. 22. Dordrecht: Springer, 2001.

———. "The *Institut du Radium* as a 'Generic Institution.'" In *Instrumentation between Science, State and Industry: Sociology of the Sciences*, edited by Bernward Joerges and Terry Shinn. Vol. 22. Dordrecht: Springer, 2001.

———, ed. *Marie Curie. Pierre Curie*. Santa Coloma de Queralt, Spain: Obrador Edèndum, Purves, 2009. A collection of primary sources by and about the Curies.

———. "The Metrology of Radioactivity." In *Instrumentation between Science, State and Industry: Sociology of the Sciences*, edited by Bernward Joerges and Terry Shinn. Vol. 22. Dordrecht: Springer, 2001.

———. "The Uses of Accumulation." In *Instrumentation between Science, State and Industry: Sociology of the Sciences*, edited by Bernward Joerges and Terry Shinn. Vol. 22. Dordrecht: Springer, 2001.

Rowell, Rebecca. *Marie Curie Advances the Study of Radioactivity*. Minneapolis: Core Library, 2016.

Rumèbe, Gérard. *Maria Sklodowska Curie et la découverte du radium*. Paris: Palais de la Découverte, 1970.

Segré, Emilio. *From X-Rays to Quarks: Modern Physicists and Their Discoveries*. San Francisco: Freeman, 1980.

Tracy, Kathleen. *Pierre and Marie Curie and the Discovery of Radium*. Hockessin, DE: Mitchell Lane, 2005.

Wirtén, Eva Hemmungs. "The Pasteurization of Marie Curie: A (Meta) Biographical Experiment." *Social Studies of Science* 45, no. 4 (August 2015): 597–610.

Wolke, Robert L. "Marie Curie's Doctoral Thesis: Prelude to a Nobel Prize." *Journal of Chemical Education* 65 (1988): 561–73.

Wolkowski, Z. W., comp. *Pierre Curie and Marie Sklodowska: The First Century of Their Impact on Human Knowledge (Le premier siècle de leur impact sur la connaissance humaine)*. Paris: Z. W. Wolkowski, 2003.

Biographical Works That Emphasize Curie's Personal Life

Although Marie Curie's life and her science are intertwined, sources tend to stress one or the other. The following sources, although discussing Curie's importance as a scientist, emphasize her personal life. This section also includes references to her family.

Allen, John. *Marie Curie*. San Diego, CA: Reference Point Press, 2016.

Bensaude-Vincent, Bernadette. "Star Scientists in a Nobelist Family: Iréne and Frédéric Joliot-Curie." In *Creative Couples in the Sciences*, edited by Helena M. Pycior, Nancy G. Slack, and Pnina G. Abir-Am, 57–71. New Brunswick, NJ: Rutgers University Press, 1996.

Bibliothèque Nationale. *Pierre et Marie Curie*. Paris: Bibliothèque Nationale, 1967.

Biquard, Pierre. *Frédéric Joliot-Curie: The Man and His Theories*. London: Souvenir Press, 1965.

Blanc, Karin. *Pierre Curie: Correspondances*. Saint-Remy-en-l'Eau, France: Editions d'Art Monelle Hayot, 2009.

Bobińska, Helena. *Marie Curie-Skłodowska*. Warsaw: Spóldzielnia "Czytelnik," 1945.

Bredin, Jean-Denis. *The Affair*. New York: George Braziller, 1986.

Brian, Denis. *Curies: A Biography of the Most Controversial Family in Science*. Hoboken, NJ: Wiley, 2005.

Bull, Angela. *Marie Curie*. London: Hamish Hamilton, 1986.

Burdowicz-Nowicka, Maria. "Nieznane materialy do dziejów Marii Skladowskiej-Curie" (Hitherto Unknown Documents Concerning the Family of Marie Curie.) *Kwartalnik Historii Nauki i Techniki* 21 (1976): 485–96.

Cano Castro, Consuelo. *Maria Curie*. Madrid: Escuela Nacional de Artes Gráficas, 1924.

Cobb, Vicki. *Marie Curie*. New York: DK, 2008.

Concasty, Marie-Louise. *Pierre et Marie Curie*. Paris: Bibliothèque Nationale, 1967.

Cotton, Eugénie. *Les Curie*. Paris: Editions Seghers, 1963.

Curie, Éve. *Madame Curie*. Translated by Vincent Sheean. New York: DaCapo Press, 1986.

Curie, Irène. "Marie Curie, ma mère." *Europe* 32 (1954): 89–121.

Damany, André. *Pierre Curie, Frédéric Joliot-Curie: origine de leur famille, généalogies*. Paris: Compagnie Littéraire-Brédys, 2006.

De Leeuw, Adèle. *Marie Curie, Woman of Genius*. Champaign, IL: Garard, 1970.

Dunn, Andrew. *Marie Curie*. New York: Bookwright, 1991.

Edison, Erin. *Marie Curie*: North Mankato, MN: Capstone Press, 2014.

Emling, Shelley. *Marie Curie and Her Daughters: The Private Lives of Science's First Family*. London: Palgrave Macmillan, 2012.

Fullick, Ann. *Marie Curie*. Chicago: Heinemann Library, 2000.

Gidel Henry. *Marie Curie*. Paris: Flammarion, 2008.

Giroud, Françoise. *Marie Curie: A Life*. Translated by Lydia Davis. New York: Holmes and Meier, 1986.

Giroux, François. *Une femme honorable*. Paris: Èditions Stock, 1987.

Goldsmith, Barbara. *Obsessive Genius: The Inner World of Marie Curie*. New York: W. W. Norton, 2005.

Goldsmith, Maurice. *Frédéric Joliot-Curie: A Biography*. London: Lawrence and Wishart, 1976.

Gormley, Beatrice. *Marie Curie: Young Scientist*. New York: Aladdin Paperbacks, 2007.

Grady, Sean M. *Marie Curie*. San Diego, CA: Lucent Books, 1992.

Healy, Nick. *Marie Curie*. Mankato, MN: Creative Education, 2004.

Henriod, Lorraine. *Marie Curie*. New York: Putnam, 1970.

Hurwic, Józef, and Gilberto Bernardini. *Maria Skłodowska Curie nel centenario della nascita (1867–1967)*. Rome: Accademia Nazionale dei Lincei, 1968.

Keller, Mollie. *Marie Curie*. New York: Watts, 1982.

Kerner, Charlotte. *Madame Curie und ihre Schwestern: Frauen die den Nobelpreis bekamen*. Weinheim, Germany: Beltz & Gelberg, 1997.

Krull, Kathleen. *Marie Curie*. New York: Viking Children's Book, 2007.

Ksoll, Peter. *Marie Curie: Mit Selbstzeugnissen und Bilddokumenten*. Presented by Peter Ksoll and Friz Vögtle. Reinbek bei Hamburg, Germany: Rowohlt, 1988.

Landa, Edward R. *Buried Treasure to Buried Waste: The Rise and Fall of the Radium Industry*. Colorado: Colorado School of Mines, 1987.

Langevin, Paul. "Pierre Curie." *Revue du mois*, July–December 1906.

Lassieur, Allison. *Marie Curie: A Scientific Pioneer*. New York: F. Watts, 2003.

Lemire, Laurent. *Marie Curie*. Paris: Perrin, 2001.

Lepscky, Ibi. *Marie Curie*. Translated by Marcel Danesi. New York: Barron's, 1993.

Linder, Greg. *Marie Curie: A Photo-Illustrated Biography*. Mankato, MN: Bridgestone Books, 1999.

Loh-Hagan, Virginia. *Marie Curie*. Ann Arbor, MI: Cherry Lake, 2018.

Madrid, Francisco. *Vida dramática de Maria Curie*. Madrid: Esperpento Ediciones Teatrales, 2016.

Margadant, Jo Burr. *Madame le Professeur*. Princeton, NJ: Princeton University Press, 1990.

Maria Sklodowska-Curie: Memorial Issue of the Polish Oncological Journal Notwotwory. Warsaw: Publishing House of the Polish Foundation of the European School of Oncology, 1998.

Marshall, Barry J. *How to Win a Nobel Prize*. Tulsa: Kane Miller, 2019.

Martin, Alexis. *Paris: Promenades dans es vingt arrondissements*. Paris: A Hennuyer, 1890.

McCormick, Lisa Wade. *Marie Curie*. New York: Children's Press, 2006.

McKown, Robin. *Marie Curie*. New York: Putnam, 1959.

McLeese, Don. *Marie Curie*. Vero Beach, FL: Rourke, 2006.

Milani, Alice. *Marie Curie: A Life of Discovery*. Minneapolis: Graphic Universe, 2019.

Miller, Connie Colwell. *Marie Curie and Radioactivity*. Mankato, MN: Capstone Press, 2007.

Milne, Catherine. "Marie Curie, Ethics and Research." In *Celebrating the 100th Anniversary of Madame Marie Sklodowska Curie's Nobel Prize in Chemistry*, edited by M. H. Chiu, P. J. Gilmer, and D. F. Treagust, 87–102. Rotterdam: Sense, 2011.

Molina, Natacha. *Maria Curie*. Madrid: Editorial Hernando, 1977.

Monteil, Claudine. *Ève Curie: l'autre fille de Pierre et Marie Curie*. Paris: Odile Jacob, 2016.

Montgomery, Mary Ann. *Marie Curie*. Englewood Cliffs, NJ: Silver Burdett Press, 1990. A children's book in the *What Made Them Great* Series.

Müller, Adolfo Simões. *Pedra mágica e a princesinha doente: pequena história de Maria Curie e da sua descoberta*. Porto: Tavares Martins, 1949.

Ogilvie Marilyn Bailey. *Marie Curie: A Biography*. Westport, CT: Greenwood Press, 2004.

Owens, Trevor. "Madame Curie above the Fold: Divergent Perspectives on Curie's Visit to the United States in the American Press." *Science Communication* 33 (September 2010): 98–119.

O'Quinn, Amy M. *Marie Curie for Kids: Her Life and Scientific Discoveries, with 21 Activities and Experiments*. Chicago: Chicago Review Press, 2017.

Pflaum, Rosalynd. *Grand Obsession: Madame Curie and Her World*. New York: Doubleday, 1989.

———. *Marie Curie and Her Daughter Irène*. Minneapolis: Lerner, 1993.

Poynter, Margaret. *Marie Curie: Discoverer of Radium*. Hillside, NJ: Enslow, 1994.

———. *Marie Curie: Genius Researcher of Radioactivity*. Berkeley Heights, NJ: Enslow, 2015.

Pycior, Helena M. "Marie Curie's 'Anti-Natural Path': Time Only for Science and Family." In *Uneasy Careers and Intimate Lives: Women in Science 1788–1979*, edited by Pnina G. Abir-Am and Dorinda Outram, 191–215.

New Brunswick, NJ: Rutgers University Press, 1987.

———. "Pierre Curie and 'His Eminent Collaborator Mme. Curie': Complementary Partners." In *Creative Couples in the Sciences*, edited by Helena M. Pycior, Nancy G. Slack, and Pnina G. Abir-Am, 39–56. New Brunswick, NJ: Rutgers University Press, 1996.

Quinn, Susan. *Marie Curie. A Life*. New York: Simon & Schuster, 1995.

Redniss, Lauren. *Radioactive: Marie and Pierre Curie, a Tale of Love and Fallout*. New York: IT Books, 2011.

Reid, Robert. *Marie Curie*. New York: Saturday Review Press, 1974.

———. *Zycie Marii Curie*. Translated by Anna Soszynska. Warsaw: Prpszusmlo o S-ka, 1997.

Regaud, Cl. "Marie Skłodowska-Curie." *Notice nécrologique*, read at the Conseil de a Fondation Curie, 24 October 1934.

Rivers, Charles, ed. *Pierre and Marie Curie: The Lives and Careers of the Science's Most Groundbreaking Couple*. Ann Arbor, MI: Charles River, 2018.

Robinson, Andrews. *Sudden Genius? The Gradual Path to Creative Breakthroughs*. Oxford: Oxford University Press, 2010.

Sabin, Louis. *Marie Curie*. Mahwah, NJ: Troll Associates, 1985.

Santella, Andrew. *Marie Curie*. Milwaukee: Almanac Library, 2001.

Schaefer, Lola M., and Wyatt Schaefer. *Marie Curie*. Mankato, MN: Capstone Press, 2005.

Senior, John E. *Marie and Pierre Curie*. Gloucestershire, UK: Sutton, 1998.

Solimeo, Silvia. *Storia, politica, scienza: l'affaire Branly-Durie*. Lecce, Italy: Milella, 2012.

Steele, Philip. *Marie Curie: The Woman Who Changed the Course of Science*. Washington, DC: National Geographic, 2006.

Steinke, Ann E. *Marie Curie and the Discovery of Radium*. New York: Baron's, 1987.

Stine, Megan. *Who Was Marie Curie?* New York: Grosset & Dunlap, 2014.

Strand, Jennifer. *Marie Curie*. Minneapolis: Abdo Zoom, 2017.

Strathern, Paul. *Curie and Radioactivity*. London: Arrow, 1998.

Tames, Richard. *Marie Curie*. London: F. Watts, 1990.

Throp, Claire. *Marie Curie*. Chicago: Heinemann Raintree, 2016.

Venezia, Mike. *Marie Curie: Scientist Who Made Glowing Discoveries*. New York: Children's Press, 2009.

Waxman, Laura Hamilton. *Marie Curie*. Minneapolis: Lerner, 2004.

Wirtén, Eva Hemmungs. *Making Marie Curie: Intellectual Property and Celebrity Culture in an Age of Information*. Chicago: University of Chicago Press, 2015.

Wolczek, Olgierd. *Maria Sklodowska-Curie*. Warsaw: Interpress, 1975.

———. *Maria Sklodowska-Curie un ihre Familie*. Leipzig: BSB Teubner, 1977.

Woznicki, Robert. *Madame Curie, Daughter of Poland*. Miami: American Institute of Polish Culture, 1983.

———. *Maria Słodowska-Curie: Her Life and Work*. Warsaw: Polonia, 1967.

Index

Académie des Sciences, 6, **11**, 12, 18, 35, 50, 52, 53, 61, 63, 65–67, 73
 See also positions, problems with, for P. Curie
alpha particles, **11–12**, 35, 58, 72
Amagat, Émile, **12**
anarchist movement, **12**
anode, **12**, 19, 70
 See also Röntgen, Wilhelm
Annales de Physique et de Chimie, x, 97
antisemitism, 12, 27
Appell, Paul, **12–13**, 21, 25, 26, 62
Ayrton, Hertha Marks, **13**, 16, 35, 53, 69, 71
 See also Curie-Langevin affair
Ayrton, W. E., **13**

Becquerel, Alexandre-Edmond, **15**
Becquerel, Antoine-Henri, x, 2, 6, **15–16**, 26, 32, 39, 40, 55, 58, 63, 71, 77, 79, 80, 88
 See also Nobel Prize for Physics, 1903, M. and P. Curie; *Researches on Radioactive Substances*
Berzelius, Jöns Jacob, **16**, 80
bicycling vacations, 5, **16**, 39, 58, 66
 See also birth and infancy of I. Joliot-Curie
birth and infancy of I. Joliot-Curie, x, 5, 6, **16**, 21, 22, 56, 66
 See also pregnancies of M. Curie
black light, **16**, 51
 See also Blondlot, Renè; N Rays
Blondlot, René, **16**, 51, 57
 See also black light
Bodichon, Barbara, 13, **16**, 35
Boguska, Bronislawa. *See* Sklodowska, Bronislawa
Boguski, Henryk, 3, **17**, 74
 See also Boguski, Józef Jerzy

Boguski, Józef Jerzy, **17**, 36
Boltwood, Bertram Borden, **17**, 36, 72, 80
 See also radium, international standard for
Borel, Émile, **17**, 73, 89
 See also Borel, Marguerite
Borel, Marguerite, **17**, 22, 89
 See also Curie-Langevin affair
Boussinesq, Joseph Valentin, **17–18**, 76
 See also Sorbonne, M. Curie as a student at the
Branley, Edouard, x, **18**, 53
Bunsen, Robert, **18**, 47

Carlough, Marguerite, **19**
 See also radium, health problems from
Carnegie, Andrew, **19**
 See also United States, M. Curie's 1921 trip to the
cathode, **19**, 21, 32, 51, 62, 70, 77, 79, 82
Chavannes, Alice, **19**, 20
 See also childhood and education of I. Joliot-Curie
Chemical News, x, 97
childhood and education of I. Joliot-Curie, 13, **19–20**, 21, 25, 56, 61–62, 77, 88–90, 96
collaborations, marital, vii, 2, 4–5, 20, 31–32, 41, 53, 54, 58, 69, 71, 85
Commune of Paris of 1871, **20**, 33
Comptes Rendus. *See* Académie des Sciences
Comstock, Anna Botsford, **20**
Comstock, John Henry, **20**
 See also Comstock, Anna Botsford
Comte, Auguste, 3, **20**, 59, 65
Coolidge, Calvin, **20**, 39, 41
Crookes, Sir William, **20–21**

INDEX

crystallography, **21**, 65
  See also Curie, Paul-Jacques; Research of P. Curie
Curie, Eugène, 6, **21**, 22, 25, 30, 33, 56, 62, 66, 96
  See also mother, M. Curie as a
Curie, Eve Denise, x, 4, 6–7, 8, 13, 19, **21**, 25, 26, 33, 34, 40, 49–50, 55, 56, 66, 73, 81, 88–89
  See also Labouisse, Eve Curie
Curie, Irène. *See* Joliot-Curie, Irène
Curie, Marie, *3, 5, 6, 8,* **21–22**, *31, 45, 58, 81, 89*
  political involvement, 3, 4, 12, 13, 54, 65–66
  relationships, 2, 4, 7, 22, 36, 50–51, 55, 93
  See also bicycling vacations; Curie-Langevin affair, Death of P. Curie; doctoral dissertation of M. Curie; duels involving M. Curie; governess, M. Curie as a; health problems of M. Curie; "Madame Sklodowska"; mother, M. Curie as a; Nobel Prize for Chemistry, 1911, M. and P. Curie and H. Becquerel; pregnancies of M. Curie; *Researches on Radioactive Substances*; Sèvres, teaching position at; Sorbonne, M. Curie as a professor at the; Sorbonne, M. Curie as a student at the; student, M. Curie as a preuniversity; United States, M. Curie's 1921 trip to the; United States, M. Curie's 1929 trip to the; World War I, experience of M. Curie in
Curie, Paul-Jacques, 5, 21, **22**, 29, 30, 33, 47, 53, 62, 63, 65
  See also family of P. Curie
Curie, Pierre, x, 4, *5,* 9, **22**
  See also death of P. Curie; early research of M. Curie; education of P. Curie; family of P.
Curiel health problems of P. Curie; positions, problems with, for P. Curie; research of P. Curie
Curie, Sophie-Claire Depouilly, x, 5–6, 16, **22**, 33, 66
  See also family of P. Curie
Curie-Langevin affair, 2, 7, 13, 17, **22–23**, 27, 31, 34, 35, 40, 43, 46, 50–51, 53, 54–55, 56, 57–58, 59, 62, 68, 72, 74, 75–76, 79, 81
  See also duels involving M. Curie; *Gil Blas; Le Temps*; Langevin, Paul

Davy, Sir Humphrey, **25**
Davy Medal, **25**, 71
death of M. Curie. *See* health problems of M. Curie
death of P. Curie, x, 13, 19, 21, 22, **25**, 26, 30, 40–41, 50, 51, 56, 61, 62, 68, 76, 77, 96
  See also Perrin, Henriette
Debierne, André-Louis, 12, **25–26**
The Delineator, 7, 20, **26**, 54
  See also United States, M. Curie's 1921 trip to the; United States, M. Curie's 1929 trip to the
Depouilly, Sophie-Claire. *See* Curie, Sophie-Claire Depouilly
Dluska, Bronislawa (Bronia), 2, 3, 4, **26**, 33, 36, 40, 53, 65, 74, 77, 80, 82
Dluski, Kazimierz, 4, **26**, 36
doctoral dissertation of M. Curie, x, 6, **26**, 32, 69, 70, 71, 95
  See also Researches on Radioactive Substances
Dorabialska, Alicja, 8, **27**
  See also health problems of M. Curie
Dreyfus, Alfred, 12, **27**
duels involving M. Curie, 7, **27**, 35, 56
  See also Curie-Langevin affair

early research of M. Curie, ix, x, 3–4, 16, **29**, 51, 75
École de Physique et Chimie Indistrielles de la Ville de Paris, x, **29**, 65
  See also early research of M. Curie; research of P. Curie
École Normale Supérieure de Jeunes Filles, x, 6, **30**, 52
École Polytechnique, **30**, 65
education of M. Curie. *See* Sorbonne, M. Curie as a student at the; student, M. Curie as a preuniversity
education of P. Curie, **30**
Einstein, Albert, 17, **30–***31*, 62, 75
Einstein, Hans Albert, **31**
Einstein, Mileva Marić, **31–32**, 53
electrometer, **32**, 70
  See also Curie, Paul-Jacques; Curie, Pierre; Research of P. Curie
electron, 11, 19, **32**, 35 55, 62, 70, 72, 77, 79

family of P. Curie, 16, 21, 22, 29, **33**, 56
Floating University, 3, 26, **33**, 54

110

INDEX

See also Comte, Auguste
fractional crystallization, 6, **33**, 36, 64, 67
Franco–Prussian War, 20, **33–34**, 76, 79
French Academy of Science. *See* Académie des Sciences
friendships, 3, 7, 9, 12, 13, 17, 19–20, 26, 30–31, 34, 50, 53, 54–55, 61–62, 66, 71, 72, 73, 81–82
Fuller, Loie, **34**, 55
See also Curie-Langevin affair

Gegner Prize, x, 11, **25**
Geiger, Hans, **35**, 54, 72
Gil Blas, 27, **35**, 56
See also duels involving M. Curie
Girton College, Cambridge, 13, 16, **35**, 54, 87
Gleditsch, Ellen, **35–36**, 68
governess, M. Curie as a, ix, 3, 4, 29, **36**, 55, 93
See also Dluska, Bronislawa; women; Zorawski family
gymnasium, **36**, 78
See also student, M. Curie as a preuniversity
Gymnasium Number Three, ix, 3, 17, **37**, 66, 70, 77, 78
See also student, M. Curie as a preuniversity

Harding, Warren, 7, *8*, 20, **39**, 82
See also United States, M. Curie's 1921 trip to the
Harvard College Observatory, **39**, 62, 71
See also Curie-Langevin affair; Joliot-Curie, Frédéric
health problems of M. Curie, 3, 4, 6–7, 8, 9, 17, 26, 27, 31, **39–40**, 53, 58, 73, 82
health problems of P. Curie, 6, 7, **40–41**, 73
See also death of P. Curie; health problems of M. Curie
Hoover, Herbert, **41**, 82
Huggins, Margaret Lindsay, **41**, 71
Huggins, William, **41**, 71

impressionism, 43
Independent, **43**
Institut du Radium. *See* Radium Institute
Institute Pasteur. *See* Pasteur Institute
intellectual climate of Paris during M. Curie's student days, **43**

International Bureau des Poids et Mesures. *See* International Bureau of Weights and Measures near Paris
International Bureau of Weights and Measures near Paris, **43**, 68
International standard for radium. *See* radium, international standard for
L'intransigeant, **43**
See also Académie des Sciences

Joachimsthal region, 6, **45**, 47, 63
See also Klaproth, Martin Heinrich
Joliot, Frédéric. *See* Joliot-Curie, Frédéric
Joliot-Curie, Frédéric, 8–9, 40, **45**, 46, 51, 53, 58, 88
Joliot-Curie, Hélène, 9, **45**, 53
See also Langevin-Joliot, Hélène
Joliot-Curie, Irène, x, 5, 6, 8, 9, 13, 16, 19–20, 21, 22, 25, 33, 40, **45–46**, 49, 50, 51, 53, 55, 56, 58, 62, 66, 77, 81, 88, 89–90, 96
See also birth and infancy of I. Joliot-Curie; childhood and education of I. Joliot-Curie; marriage of Irène Curie to Frédéric Joliot; Nobel Prize for Chemistry, 1935, I. Joliot-Curie and F. Joliot-Curie; World War II, experience of I. Joliot-Curie in
Joliot-Curie, Jean Frédéric. *See* Joliot-Curie, Frédéric
Joliot-Curie, Pierre, **46**
Le Journal, 23, **46**, 57, 76, 79
See also Curie-Langevin affair

Kelvin, Lord, **47**, 69
Kirchoff, Gustav Robert, 18, 41, **47**
Klaproth, Martin Heinrich, **47**, 83

laboratory, ix, *1*, 2, 4, 7, 8, 9, 16, 17, 29, 30, 33, 36, 40, 45, 51, 52, 53, 55, 56, 65, 67, 68, 69, 75, 76, 82, 85, 89
Labouisse, Eve Denise Curie, x, 4, 6–7, 8, 13, 19, 21, 25, 26, 33, 34, 40, **49–50**, 55, 56, 66, 73, 81, 88–89
Lamotte, 4, **50**
Langevin, Emma Jeanne Defosses, 22–23, 31, 46, **50**, 51, 57, 59
See also Curie-Langevin affair; duels involving M. Curie
Langevin, Paul, 2, 7, 17, 20, 22–23, 27, 45, 46, **50–51**, 73, 75

See also Curie-Langevin affair; Langevin, Emma Jeanne Defosses
Langevin-Joliot, Hélène, 45, **51**
  See also Curie-Langevin affair
Lankester, E. Ray, **51**
Lankester, H. Edwin, **51**
  See also Lankester, E. Ray
Le Châtelier, Henri, 29, **51**
  See also early research of M. Curie
Lebon, Gustave, 16, **51**
Lenard, Philipp, **51**, 62, 70
Leslie, May Sybil, **51**
  See also Gleditsch, Ellen; Joliot-Curie, Irène; Perey, Marguerite; Ramstedt, Eva Julia Augusta
leukemia, **52**
  See also radium, health problems from
Lippmann, Gabriel, **52**, 76
Lodge, Sir Oliver, **52**

"Madame Sklodowska", 13, **53**
  See also Curie-Langevin affair; health problems of M. Curie
Marconi, Guglielmo, 18, **53**
Marić, Mileva, 31, **53**
  See also Einstein, Mileva Marić
marriage of Irène Curie to Frédéric Joliot, 45, **53**
Marsden, Ernest, 35, **54**, 72
Marxism, 33, **54**
Maunder, Annie Scott Dill Russell, **54**
Maunder, Walter, **54**
Meloney, Marie Mattingly (Missy), x, 7, 9, 20, 26, 39, **54–55**, 81–82
  See also United States, M. Curie's 1921 trip to the
Meyer, Stefan, **55**
  See also radium, international standard for
Michalowska, Henrietta, **55**
  See also governess, M. Curie as a
Millikan, Robert, **55**
Mittag-Leffler, Gösta, **55–56**
Mortier, Pierre, 27, **56**
  See also Curie-Langevin affair
mother, M. Curie as a, 5, 16, 39, 49, **56**, 57, 90, 96
Mouton, Henri, **56**

N rays, 16, 51, **57**
Neumann, Elsa, **57**
  See also women

New York Times, 23, 31, 50, 55, **57**, 82
  See also Curie-Langevin affair
Nobel Prize for Chemistry, 1911, M. Curie, x, 2, 7, **57–58**, 77, 88
  See also Curie-Langevin affair; women
Nobel Prize for Chemistry, 1935, I. Joliot-Curie and F. Joliot-Curie, 9, 45, 53, 63, **58**
Nobel Prize for Physics, 1903, M. and P. Curie and H. Becquerel, x, 2, 6, 15, 39, 55, *58*, 77, 88

L'Oeuvre, 23, 27, 56, 57, **59**
Orzeszkowa, Eliza, **59**

Pantheon of Paris, x, 9, **61**
Pasteur Institute, 7, 8, 56, **61**, 68, 71
  See also Radium Institute
Pegram, Dean George, *81*
Perey, Marguerite, 11, **61**
  See also women
Perrin, Aline, **61–62**
Perrin, Francis, 61, **62**
Perrin, Henriette Jeanne Eugenie Blanche Duportal, 17, 20, 25, 61, **62**, 89
Perrin, Jean, 17, 20, 21, 25, 61, **62**, 89
  See also death of P. Curie
phosphorescence, 15–16, 32, **62**, 63, 79
Pickering, Edward, 39, **62**
piezoelectric quartz balance, 21, 29, **62–63**, 70
  See also crystallography
piezoelectricity, 21, 29, 47, 50, 52, 62, **63**, 65
  See also crystallography; piezoelectric quartz balance
pitchblende, 6, 16, 18, 33, 45, 47, **63**, 64, 67, 80, 83
plum pudding model of atom. *See* Geiger, Hans
Poincaré, Henri, 15, **63**, 76
  See also positions, problems with, for P. Curie
Poland, ix, x, 2–4, 9, 20, 22, 26, 33, 49, 54, **63–64**, 65, 66, 68, 71–72, 74, 76, 77, 82, 87–88, 89, 99
polonium, x, 1, 6, 12, 20, 47, 57, 63, **64**, 65, 67, 69
positions, problems with, for P. Curie, 6, 29, 30, 62, 63, **64–65**
  See also research of P. Curie

INDEX

positivism, 12, 20, 33, 54, 59, **65–66**
pregnancies of M. Curie, x, 5–7, 16, 21, 58, **66**
Prix la Caze of the Academy of Sciences, 11, **66**
Przyborovska, Kazia, 3, **66**

radioactivity, vii, x, 1, 5, 7, 8, 9, 12, 16, 17, 19, 26, 32, 33, 36, 40, 45, 55, 58, 61, 64, **67**, 69, 70, 73, 75, 80, 97, 99
radioactivity and cancer. *See* radium, health problems from; leukemia
radioactivity and cataracts. *See* radium, health problems from
radium, x, 1, 6–8, 9, 11, 12, 13, 17, 19, 20, 25, 26, 27, 34, 36, 39–40, 41, 43, 47, 50, 52, 54, 55, 57, 61, 63, 64, 65, **67**, 68, 69, 70, 73, 81–82, 85, 89
   as a cure, 7, 55, 69
   health problems from, 6–8, 19, 27, 39–41, 52, 58, 66, **67–68**, 69, 73
   international standard for, 17, 36, 43, 55, **68**
   See also health problems of M. Curie; health problems of P. Curie; Meloney, Marie Mattingly; polonium; radioactivity
Radium Institute, 7, 26, 27, 34, 61, **68–69**, 71, 88
   See also Curie-Langevin affair
radon gas, 51, 67, **69**, 72
Ramsay, Sir William, 36, **69**
Ramstedt, Eva Julia Augusta, **69**
   See also Gleditsch, Ellen; Joliot-Curie, Irène; Leslie, May Sybil; Perey, Marguerite
Rayleigh, Lord, **69**, 77
Recherches sur les Substances Radioactives, **69**
   See also Researches on Radioactive Substances
Regaud, Claudius, 8, 61, 68, **69–70**
   See also Pasteur Institute; Radium institute
religion, 3, 5
research of P. Curie, 4–6, **70**
Researches on Radioactive Substances, 69, **70**
   See also Recherches sur les Substances Radioactives
Rogowska, Anna Maria, **70**
   See also Rogowski, Zdzislaw
Rogowski, Zdzislaw, **70**
Röntgen, Wilhelm, 15, 16, 19, 32, 51, 62, 63, **70–71**, 73, 82
   See also phosphorescence

Roux, Pierre Paul Émile, 8, 61, **71**
Royal Greenwich Observatory, 54, **71**
Royal Institution, London, 6, 25, 40, **71**
Royal Society of London, 11, 13, 25, 50, 51, **71**, 77, 79
Russia, 2–3, 17, 26, 33, 37, 63–64, 66, **71–72**, 73, 74, 75, 76, 77–78, 80, 82, 87, 88, 89
Rutherford, Ernest, First Baron Rutherford of Nelson, 11–12, 17, 35, 52, 54, 55, 69, **72**, 75, 79, 80

Sagnac, Georges, **73**
Sancellemoz Sanatorium, 9, 40, **73**
   See also health problems of M. Curie
Schmidt, Gerhard Carl, 12, **73**
Sèveres. *See* École Normale Supérieure de Jeunes Filles
Sèveres, teaching position at. *See* École Normale Supérieure de Jeunes Filles
Sikorska, Jadwiga, ix, 2, 3, 37, **73**, 74, 77, 80
   See also student, M. Curie as a preuniversity
Sklowdowska, Bronislawa (Bronia). *See* Dluska, Bronislawa (Bronia)
Sklodowska, Bronislawa, ix, 2–3, 17, 22, 37, **73–74**, 75, 77, 80
Sklodowska, Helena (Hela), 2, 26, 73, **74**, 75, 77, 80
Sklodowska, Józef, 2, 3, 26, 37, **74–75**
Sklodowska, Maria (Manya). *See* Curie, Marie
Sklodowska, Zofia (Zosia), ix, 2, 3, 17, **74**, 80
Sklodowski, Wladyslaw, ix, 2, 22, 73–74, **75**, 80, 82
Society for the Encouragement of National Industry, ix, 29, **75**
   See also early research of M. Curie
Soddy, Frederick, 12, **75**
Solvay, Ernest, **75**
Solvay Conferences, 22–23, 31, 46, **75–76**
   See also Curie-Langevin affair; Solvay, Ernest
Sorbonne, ix, x, 4, 6, 7, 8, 12–13, 15, 17, 20, 26, 29, 30, 34, 36, 50, 51, 52, 56, 57, 61, 62, 63, 65, 68, 70, 71, **76**, 77, 79, 87, 97
   M. Curie as a professor at the, 7, 30, **76**
   M. Curie as a student at the, 6, 17, **76–77**
spiritualism, 21, 52
Stockholm, 7, 53, 58, **77**
   See also Nobel Prize for chemistry, 1911, M. Curie; Nobel Prize for physics, 1903, M. and P. Curie and H. Becquerel

INDEX

Stokes, George, **77**
Stoney, George Johnstone, **77**, 79
Strutt, John William. *See* Rayleigh, Lord
student, M. Curie as a preuniversity, ix–x, 2–3, 4, **77–78**
Szalay, Helena (Hela). *See* Sklodowska, Helena (Hela)

Le Temps, **79**
Téry, Gustave, 27, 56, 59, **79**
  See also Curie-Langevin affair
Third Republic, 12, 20, 27, 33, **79**
  See also M. Curie as a student at the
Thompson, Silvanus P., 77, **79**
Thompson, Joseph John, **79–80**
Thompson, William. *See* Kelvin, Lord
thorium, 12, 16, 51, 69, 73, **80**
Treatise on Radioactivity, **80**
Tupalska, Antonina, **80**
  See also student, M. Curie as a preuniversity
typhus, ix, 3, 17, 74, **80**

United States, M. Curie's 1921 trip to the, 7, 39, **81–82**
  See also The Delineator
United States, M. Curie's 1929 trip to the, 9, 41, **82**
University of St. Petersburg, 75, **82**
University of Würzburg, **82**
uranium, 6, 15–16, 17, 26, 36, 45, 47, 63, 64, 67, 69, 70, 79, 80, **83**

Vassar College, **85**, 87
Villard, Paul, **85**
Vogt, Oskar, **85**
Vogt-Mugnier, Cécile, **85**

Warsaw, ix, 2–4, 9, 17, 20, 22, 26, 27, 36, 50, 64, 65, 66, 68, 71, 73, 74, 75, 77–78, 82, **87**, 93–94
women, vii–viii, 1–2, 3, 5–6, 7, 8, 11, 13, 16, 18, 20, 27, 30, 33, 35–36, 39, 41, 43, 51, 52, 53, 54, 55–56, 57, 58, 61, 62, 69, 71, 76, 81–82, 85, **87–88**, 90
women Nobel Prize laureates, x, 2, 6, 7, 9, 23, 34, 39, 45, 50, 52, 53, 55–56, 57–58, 66, 77, **88**
women scientists in Marie Curie's laboratory. *See* Gleditsch, Ellen; Joliot-Curie, Irène; Leslie, May Sybil; Perey, Marguerite; Ramstedt, Eva Julia Augusta
World War I, experience of I. Joliot-Curie in, **88**, 89–90
World War I, experience of M. Curie in, 7, **88–90**
World War II, experience of I. Joliot-Curie in, **90**

X-rays. *See* Röntgen, Wilhelm

Zorawski family, ix, 3–4, 36, **93**
Zorawski, Kazimierz, 4, 36, 55, 66, **93–94**
  See also Zorawski family

About the Author

Marilyn Bailey Ogilvie is professor emerita of the history of science and curator emerita of the History of Science Collections at the University of Oklahoma. She was born in Duncan, Oklahoma, and grew up in Kansas City, Missouri. She has three children and six grandchildren. She received her B.A. degree in biology from Baker University, Baldwin, Kansas, and her M.A. degree in zoology from the University of Kansas, Lawrence, Kansas. She taught in Phoenix, Arizona, for two years and then went with her husband to East Africa to teach for a State Department program, Teachers for East Africa, for two years. Upon returning to the United States with her six-week-old daughter, she studied the history of science at the University of Oklahoma, Norman, Oklahoma, and received her Ph.D. degree from that institution. Her dissertation was on the various editions of the pre-Darwinian evolutionary theorist Robert Chambers's work *Vestiges of the Natural History of Creation*. She also earned an M.A. degree in library science from the University of Oklahoma. She has taught at Portland State University, St. Andrew's College, Minaki (now Tanzania East Africa), Oklahoma Baptist University, and the University of Oklahoma. She became curator of the History of Science Collections at OU from 1991 to 1994, when she became emerita. During this time, she also taught in the History of Science Department and mentored graduate students. As curator, she traveled extensively in Europe during the summers to purchase rare books in the history of science.

She was a distinguished lecturer (2012) and was a member of the Rossiter Prize Committee for the History of Science Society (2011). She received the distinguished teaching award for 1989 at Oklahoma Baptist University (where she was chair of the Division of Natural Science and Mathematics) and the Outstanding Faculty Award from the University of Oklahoma Student Association. She was the editor for *Landmarks of Science*, project coordinator and academic humanist for the Oklahoma Humanities Committee for seven projects, peer reviewer for NEH and NSF, honorary Phi Beta Kappa (1993), and a member of Omicron Delta Kappa. As curator of the History of Science Collections, she received two grants from the Mellon Foundation and two from the Department of Education. Although she was interested in all aspects of the history of science, she became especially fascinated by the history of women in science. She published over 50 papers and chapters in books, reviewed the works of others, and gave many lectures on this subject.

Her previous books include *Women in Science: Antiquity through the Nineteenth Century; A Biographical Dictionary with Annotated Bibliography* (1986), *Women and Science: An Annotated Bibliography* (1996), *A Dame Full of Vim and Vigor: A Biography of Alice Middleton Boring, an American Biologist in China* (with Clifford J. Choquette, 1999), *Biographical Dictionary of Women in Science* (2 vols., ed. with Joy Harvey, 2000), *Marie Curie: A Biography* (2004), *Sweeping the Stars: The Story of Caroline Herschel* (2008), and *For the Birds: American Ornithologist Margaret Morse Nice* (2018).